Electrospinning - Theory, Applications, and Update Challenges

Edited by Khalid S. Essa
and Khaled H. Mahmoud

Published in London, United Kingdom

Electrospinning - Theory, Applications, and Update Challenges
http://dx.doi.org/10.5772/intechopen.107552
Edited by Khalid S. Essa and Khaled H. Mahmoud

Contributors
Abdul Ghani Razaqpur, Cláudia Mouro, Guillermo M. Herrera-Pérez, Isabel C. Gouveia, Khaled H. Mahmoud, Khalid S. Essa, Manuel F. Piñón-Espitia, Matha T. Ochoa-Lara, Sharvare Palwai, Xiaocheng Zhang, Yuan Liao

First published in London, United Kingdom, 2024 by IntechOpen
IntechOpen is the global imprint of INTECHOPEN LIMITED, registered in England and Wales, registration number: 11086078, 5 Princes Gate Court, London, SW7 2QJ, United Kingdom

British Library Cataloguing-in-Publication Data
A catalogue record for this book is available from the British Library

Additional hard and PDF copies can be obtained from orders@intechopen.com

Electrospinning - Theory, Applications, and Update Challenges
Edited by Khalid S. Essa and Khaled H. Mahmoud
p. cm.
Print ISBN 978-1-80355-670-3
Online ISBN 978-1-80355-671-0
eBook (PDF) ISBN 978-1-80356-152-3

For EU product safety concerns:
IN TECH d.o.o., Prolaz Marije Krucifikse Kozulić 3, 51000 Rijeka, Croatia,
info@intechopen.com or visit our website at intechopen.com.

Meet the editor

Dr. Khalid S. Essa received a BSc (with honors), MSc, and Ph.D. in Geophysics from the Faculty of Science, Cairo University, Egypt, in 1997, 2001, and 2004, respectively. He joined the staff of Cairo University in 1997 and was appointed Research Professor of Potential Field Methods in the Department of Geophysics in 2014. He completed post-doctoral studies at Strasbourg University, France; Charles University Prague, Czech Republic; and Western Michigan University, USA. He has served as an editor and external reviewer for many top journals. He attended several international geophysical conferences in the United States, Australia, and France. He is a member of the Society of Exploration Geophysicists (SEG), the American Geophysical Union (AGU), the American Association of Petroleum Geologists (AAPG), the European and Association of Geoscientists & Engineers (EAGE). He is also a member of the Petroleum and Mineral Resources Research Council, Academy of Scientific Research and Technology, Egypt. Dr. Essa is the recipient of the 2017 Prof. Nasry Matari Shokry Award in Applied Geology from the Academy of Scientific Research & Technology and an award from Cairo University for Scientific Excellence in Interdisciplinary, Multidisciplinary and Future Sciences in 2017.

Khaled Hussein Mahmoud, Ph.D., is a professor in the Department of Physics, at Khurmah University College, Taif University, KSA. He works in the field of polymers and polymer nanocomposites. His main research lies in studying the electrical, optical, and thermal properties of polymers. Dr. Mahmoud studied physics at Cairo University, Egypt. He also completed a professorship at the University of Cairo. He now works at Taif University, KSA.

Contents

Preface

Electrospinning, a fascinating and versatile technique, has revolutionized the field of nanotechnology and materials science. With its ability to produce ultrafine fibers, electrospinning has found applications in diverse areas such as tissue engineering, drug delivery, filtration, and electronics. This book provides a comprehensive overview of electrospinning, covering its principles, methods, applications, and prospects.

The first section of this book consists of three chapters that explore topics ranging from the fundamental principles of electrospinning to the latest advancements in the field. Each chapter is meticulously crafted to provide a blend of theoretical insights and practical applications, making it a valuable resource for researchers, engineers, and students alike.

The second section of the book contains two chapters that present a clear picture of applications in electrospinning. It examines the intricacies of fiber formation, the influence of process parameters, and the wide array of materials that can be spun into fibers. Moreover, the chapters explore the exciting possibilities that electrospun fibers offer in novelty style.

We hope that this book will not only serve as a guide for researchers and practitioners but also inspire future innovations in this exciting field. We invite you to embark on this journey with us and explore the limitless possibilities of electrospinning.

Khalid S. Essa, Ph.D.
Professor,
Faculty of Science,
Geophysics Department,
Cairo University,
Egypt

Khaled H. Mahmoud, Ph.D.
Professor,
Department of Physics,
Khurma University College,
Taif University,
Taif, Saudi Arabia

Section 1

Electrospinning Background and Revision

Chapter 1

Introductory Chapter: The Electrospinning

Khaled H. Mahmoud and Khalid S. Essa

1. Introduction

The production of nanofibers from a polymer solution or melt may be accomplished using an electric field called electrospinning. Due to the peculiar characteristics of the nanofibers produced by this procedure, there has been a significant interest in various areas, including medicine, biotechnology, and materials science [1]. Originally conceived as a theoretical concept, electrospinning has evolved into a versatile and widely used process with diverse applications across several sectors. Due to its ability to produce nanofibers with exact control over their structure and composition, this method benefits researchers and industries [2].

2. History of electrospinning

L.J. Cohn and J.A. Sprague first proposed the process of electrospinning in 1934, and they also secured a patent for it. At the same time, Formhals also obtained a separate patent for a method that closely resembled electrospinning [3, 4]. However, the first applications and understanding of electrospinning were limited, and the technique did not get much attention from the research community at that time. Taylor's work in the 1960s had a significant role in the progress of electrospinning by developing a model for the hopper configuration, where an electric field produces liquid droplets [5, 6]. His partnership with JR Melcher resulted in the creation of the "Leaky Dielectric Model" for conductive liquids [7].

Electrospinning gained practical importance in the 1990s as a consequence of researchers creating more controlled techniques and beginning to investigate applications in a range of sectors [8, 9]. In particular, Darrel Reneker and Gregory were able to effectively decrease a variety of organic polymers to the nanoscale by the use of electrospinning [10, 11]. In order to manufacture ultrafine fibers with dimensions that are smaller than 5 micrometers, Darrel Reneker was the pioneer who introduced the method of submitting a polymer dispersion to high voltage levels. This was the beginning of the technology's rise to prominence. In addition, during this time period, advancements in polymer science and materials engineering led to the creation of electrospinning as a method that had the potential to be regarded as practical [12].

Electrospinning is an intriguing technique used to produce fibers with advantageous structural characteristics, as seen in **Figure 1**.

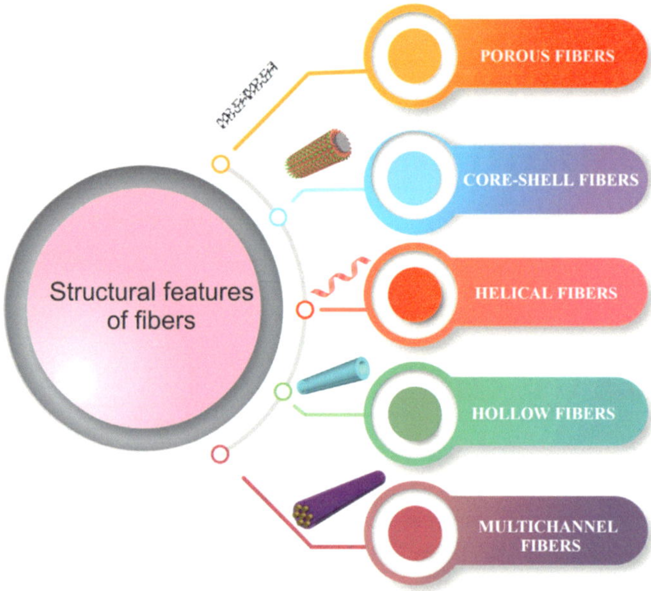

Figure 1.
Different types of fibers synthesized via *electrospinning method.*

3. Fibers manufactured by electrospinning

3.1 Porous fibers

These fibers are known as porous fibers because they have holes or empty areas within their structure. The size and form of the pores further define their practical applicability, but in general, porous fibers are used in filtering because of their capacity to trap particles [13].

3.2 Core-shell fibers

The core of these fibers is made of a distinct substance, while the shell is made of a different material. Due to the fact that each component may contribute its own unique traits, this structure makes it possible to create one-of-a-kind combinations of attributes [14].

3.3 Helical fibers

A twisted or coiled structure is characteristic of fibers that are spiral or helical in form. It is possible for the fibers to acquire intriguing mechanical and optical characteristics as a result of this winding [15].

3.4 Hollow fibers

The structure of hollow fibers is characterized by the presence of voids or empty areas. The empty area may be filled with a particular material, making this design ideal for applications, such as medicine administration, where it can be filled with specific substances [16].

3.5 Multichannel fibers

Multi-channel fibers, often known as fibers having several routes or channels, provide unique capabilities that are not found in other fiber kinds. Due to their versatility, multichannel fibers provide a promising and adaptable foundation for many technological applications. This is especially accurate in fields where meticulous regulation of fluid flow or sensing is crucial. Researchers persist in exploring and developing innovative uses for these fibers in many applications and areas of research [17].

4. Basic electrospinning methods

There are a few distinct methods that are used in the manufacturing of fibers from polymers. These methods include wet spinning, dry spinning, liquefying, and gel spinning. Every approach has its own set of qualities, benefits, and applications that are unique to him. Following is a summary of the distinctions that exist between these various spinning processes:

4.1 Wet-spinning

Extruding a polymer solution through a spinneret and into a coagulation bath is the process involved in wet spinning. It is the presence of a non-solvent for the polymer in the coagulation solution that causes the polymer to solidify and form fibers. The use of wet spinning is advantageous because it may be used for polymers that are difficult to dissolve in conventional solvents. The manufacture of fibers with high strength is made possible as a result. Wet spinning is often used in the production of rayon and acrylic fibers, among other applications [18].

4.2 Dry spinning

The process of dry spinning involves the extrusion of a polymer solution into a heated air chamber by use of a spinneret. The fibers become hardened when the solvent evaporates, leaving behind the fibers. In the case of polymers that are capable of being dissolved in volatile solvents, dry spinning is an advantageous technique. It enables the manufacture of fibers that have certain qualities, such as porosity, among other characteristics. In the context of applications, dry spinning is often used for the production of acetate fibers and some kinds of polyesters [19].

4.3 Melt spinning

Melting a polymer and then extruding it *via* a spinneret is the process that is involved in the liquefying process, which is also referred to as melt spinning. When the polymer is allowed to cool, it becomes solid and forms fibers. A number of advantages include the fact that liquefying is appropriate for thermoplastic polymers that may be melted without suffering any chemical transformations. This is a basic procedure that has relatively minimal costs associated with manufacturing.

Molten spinning is often used in the production of nylon, polyester, and polypropylene fibers, among other applications [20].

4.4 Gel spinning

Gel spinning is a process that includes the creation of a gel from a polymer solution or melt. The gel generated by this process is then pulled or stretched in order to align the polymer chains. High-strength fibers are formed by the gel that has been aligned and then hardened. Gel spinning results in fibers that have good strength and modulus, which is a significant advantage. It is possible to improve the mechanical characteristics of the material by aligning the polymer chains during the drawing process. The production of high-performance fibers such as aramid fibers (e.g., Kevlar) is often accomplished by the use of gel spinning [21].

5. Variation in electrospinning

Electrospinning may be categorized into various variations and types, each designed for specific purposes or desired outcomes. The selection of the electrospinning technology is based on the desired characteristics of the nanofibers, including their composition, structure, shape, and specific parameters relevant to the intended application.

5.1 Conventional electrospinning

This is the most basic kind of electrospinning, in which a polymer solution or melt is delivered to a spinneret using a syringe pump. A high voltage is delivered between the spinneret and a grounded collector, causing nanofibers to develop [22].

5.2 Coaxial electrospinning

The process of coaxial electrospinning includes the simultaneous spinning of several fluids with the use of needles that are either concentric or coaxial. Through the use of this technique, it is possible to create core-shell structures, in which one component acts as the core and another substance encloses it as a shell [23].

5.3 Needleless electrospinning

Needleless electrospinning employs a spinneret-free setup, such as a rotating disk or drum, instead of the conventional spinneret utilized in the process. This approach is highly suitable for large-scale production and has the capacity to streamline the electrospinning process [24].

5.4 Emulsion electrospinning

When one liquid is suspended in another in the form of tiny droplets, the result is an emulsion. One method for creating droplets within a continuous phase is emulsion electrospinning. This technique makes use of emulsions, which are combinations of two liquids that do not readily blend together. As the liquid evaporates, the electric field induces the formation of fibers [25].

5.5 Rotary jet-spinning

The production of nanofibers is accomplished by the use of a spinning spinneret in this method. A centrifugal force is generated as a result of the spinneret's spinning, which makes it possible to deposit fibers on a collector in a regulated manner [26].

5.6 Bubble electrospinning

The electrospinning process is altered by the incorporation of a gas using the bubble electrospinning technology. Depending on the gas and polymer solution mixture, this process might potentially lead to the formation of porous structures or hollow fibers [27].

5.7 Melt electrospinning

The use of molten polymer rather than a polymer solution is what is involved in the process of melt electrospinning. As the molten polymer reaches the collector, it undergoes the process of electrospinning, which results in the formation of fibers [28].

5.8 Side-by-side electrospinning

Side-by-side electrospinning is a technique that involves electrospinning two or more polymer solutions concurrently *via* separate channels, yet the resulting solutions are collected together. This enables the development of composite fibers that include components that are unique from one another [29].

5.9 Multi-jet electrospinning

A multi-jet electrospinning setup uses a number of nozzles or jets to electrospin several polymer solutions simultaneously. Complex fiber structures may be created using this technique [30].

6. Electrospinning process configuration

An electrospinning setup consists of three primary components: (i) a voltage source, (ii) a metal fiber, and (iii) a semiconductor collector [31]. Both of these components are essential to the process. It is possible to further subdivide the semiconductor collector into three primary groups, which are as follows: (a) fixed flat plate collectors, (b) rotating drum collectors, and (c) rotating disk collectors [32].

7. Working principle

At the location of the liquid droplets created at the needle's tip, an adequate voltage is applied. The liquid gets charged. Electrostatic repulsion counteracts the surface tension. Thus, the droplet is stretched, and at a critical point, a stream of liquid bursts from the surface and spirals into a cone-like structure termed a "Taylor cone." Once the Taylor cone is constructed, the fluid jet is directed to the metal collector. The formation of solid fibers may be attributed to the liquid's thickness, cooling, or evaporation. The solvent's swirling action causes the Taylor cone to evaporate off the collector during flight, coating the collector with a substance other than fiber [33–35].

8. Parameter optimizations for the electrospinning process

The quality of electrospun fibers generated from polymer solutions may be influenced by critical parameters, which can be classified into three subclasses as given in **Figure 2** [36]:

1. Solution parameters

2. Process parameters

3. Environmental parameters

8.1 Solution parameters

A lot of things about the solution affect the properties of electrospun fibers made from polymer solutions. These include how fast the solvent evaporates, how thick the polymer solution is, how concentrated it is, and how high the surface tension is [37]. The concentration and viscosity of the produced fiber have a direct proportional relationship and have comparable effects. The creation of a bead is attributed to a polymer solution with low viscosity or low concentration, while a polymer solution with greater viscosity and concentration results in fibers with a larger diameter. Typically, raising the polymer solution concentration leads to a larger fiber diameter. The polymer solution exhibits bead formation due to its elevated surface tension, whereas a decrease in surface tension promotes the creation of smooth fibers. The polymer's concentration, surface tension, and solvent viscosity influence the electrons' spin rate and the fibers' shape [38].

8.2 Process parameters

The surface tension of the polymer solution has a significant impact on the critical voltage, which is the voltage at which the charged jet begins the electrospinning process. The diameter of the electrospun fibers exhibited an inverse relationship with the applied voltage, such that lowering the voltage resulted in an increase in diameter, and

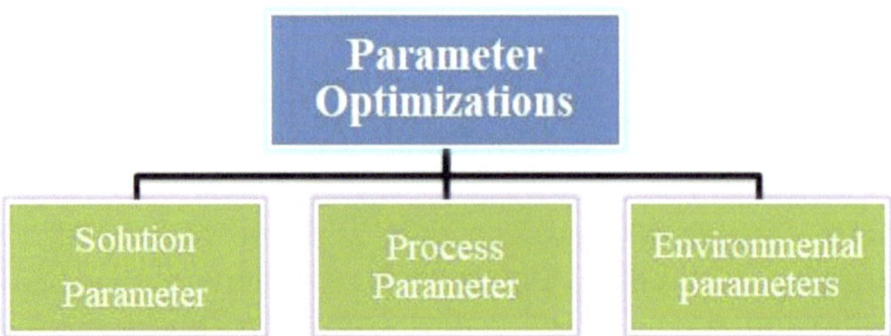

Figure 2.
Optimizing parameters for electrospinning of fibers.

vice versa. The primary factor influencing fiber solidification is the distance between the tip and the collector since it determines the amount of time the threads have to dry before reaching the collector. An insufficient gap encourages the development of beads, while a greater length yields bead-free fibers [39].

8.3 Environmental parameters

Humidity and temperature are regarded as factors that are distinctive to the environment. The characteristics of electrospun fibers are influenced indirectly by these factors. By raising the temperature, the rate of evaporation rises, leading to an elevation in viscosity and concentration. Consequently, this promotes the creation of fibers with a greater diameter. Furthermore, the viscosity of the solution often reduces as the temperature rises. With an increase in humidity, there is a corresponding rise in the average diameter of the fibers [40].

9. Applications of electrospun nanofibers

Electrospun nanofibers have been used in a broad variety of sectors as a result of their exceptional characteristics, which include a large surface area, a tiny diameter, and a shape that can be adjusted. The many domains in which electrospun nanofibers are used are outlined in **Figure 3**.

Electrospun nanofibers have a number of significant uses some notable applications of electrospun nanofibers include:

9.1 Biomedical

Electrospun nanofibers have the potential to serve as drug delivery systems, offering precise and regulated release of medications while enhancing their bioavailability. Nanofibrous scaffolds are made to look like the extracellular matrix. This makes it easier for cells to attach and for tissues to grow back in areas like bone tissue engineering and wound healing. Electrospun nanofibers may be used to generate wound

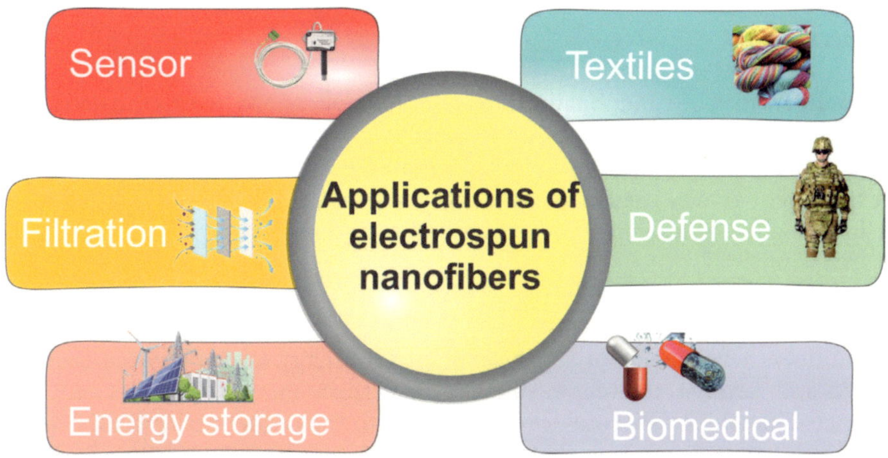

Figure 3.
Applications of electrospun nanofibers.

dressings that are both permeable and very porous. These dressings help encourage healing by maintaining a moist environment and avoiding bacterial infection.

9.2 Textile

Nanofibers may be introduced into textiles to offer new capabilities, such as resistance to water, antibacterial characteristics, and better breathability. These functionalities can further enhance the textiles' overall performance. Electrospun nanofibers have the potential to improve the protective characteristics of clothing, making it more resistant to the effects of environmental conditions, diseases, and toxins.

9.3 Defense

Electrospun nanofibers' unique qualities make them useful in defense and military technology. Ballistic textiles made using nanofibers are lightweight, flexible, and projectile-resistant. Electrospun nanofibers may be functionalized to make chemical- and biological-resistant garments. Gas filtration using nanofibrous membranes protects military troops from harmful gasses and chemical warfare weapons. Electrospun nanofibers enhance air quality in confined military areas by filtering particulates. Nanofibers can make lightweight, flexible communication antennas. They may be used as sensors to provide real-time environmental data. Electronic warfare methods like signal jamming and interference may use electrospun nanofibers. Nanofibers having radar-absorbing or deflecting qualities may be used in military stealth equipment. Electrospun nanofibers may be utilized to make antibacterial wound dressings for speedier field injury recovery.

9.4 Filtration

Nanofibrous membranes are well-suited for air and water filtration applications due to their expansive surface area and minuscule pore size. These membranes have the ability to extract particles and impurities from both the air and water. Electrospun nanofibers may be used for the specific separation of oil and water, making them potentially valuable in environmental cleanup and industrial applications.

9.5 Sensor

Nanofibers may be functionalized for sensing applications, which can detect changes in temperature, humidity, or certain chemical compounds. This technology applies to the field of sensor technology. Flexible Electronics: Electrospun nanofibers have the potential to be used as components in electronic devices that are both flexible and lightweight. These devices include capacitors and sensors.

9.6 Energy storage

Nanofibers have the potential to be exploited in energy storage devices, such as fuel cells and batteries, in order to enhance the performance of electrodes and the efficiency with which they conduct energy. Nanofibrous materials synthesized via the electrospinning approach have the potential to increase the efficiency of solar cells by offering a large surface area for the absorption of light and by enhancing the movement of electrons.

Nanofibers that have been electrospun are becoming more versatile, and current research efforts are helping to broaden their uses across a variety of sectors. Because they may be modified in terms of their qualities and usefulness, they are very useful in tackling unique difficulties that are present in a variety of sectors.

10. Future prospects

Electrospinning is a method for creating a wide range of one-dimensional materials with adjustable structural properties. A lot of ground has been covered, but we still have a ways to go. Attaining consistent size and shape in nanofibers, while simultaneously satisfying the demands of end-users continues to pose a significant challenge. There are fewer diverse uses for natural polymers because they lack the mechanical and chemical characteristics of synthetic polymers. Researchers are working on hybrid polymer systems that include synthetic and natural polymers to address this. These systems are electrospun and provide improved functionality, particularly in biotechnology. In tissue engineering, adding nanoparticles after electrospinning therapy will increase its effectiveness. Studying the interactions between nanofiber designs and various drug-related release patterns is crucial. There are many unanswered questions and untapped potential applications for electrospun nanofibers in the energy sector, in filtering applications, and as sensors. To be useful, nanofiber membranes need upgrades to their surface area, pore diameters, and other properties.

Author details

Khaled H. Mahmoud[1] and Khalid S. Essa[2]*

1 Department of Physics, College of Khurma University College, Taif University, Taif, Saudi Arabia

2 Faculty of Science, Geophysics Department, Cairo University, Giza, Egypt

*Address all correspondence to: khalid_sa_essa@cu.edu.eg

References

[1] Mingjun C, Youchen Z, Haoyi L, Xiangnan L, Yumei D, Bubakir MM, et al. An example of industrialization of melt electrospinning: Polymer melt differential electrospinning. Advanced Industrial and Engineering Polymer Research. 2019;**2**:110-115

[2] Xue J, Wu T, Dai Y, Xia Y. Electrospinning and electrospun nanofibers: Methods, materials, and applications. Chemical Reviews. 2019;**119**:5298-5415

[3] Figen AK. History, basics, and parameters of electrospinning technique. In: Electrospun Materials and Their Allied Applications. Hoboken: Wiley; 2020. pp. 53-69

[4] Kleivaitė V, Milašius R. Electrospinning–100 years of investigations and still open questions of web structure estimination. AUTEX Research Journal. 2018;**18**:398-404

[5] Fallahiarezoudar E, Yusof NM, Idris A, Ngadiman NHA. An introduction to electrospun nanofibers for tissue engineering heart valves. Nanotechnology Research Journal. 2017;**10**:423-448

[6] Pourdeyhimi B, Fedorova N, Maze B. Micro and nanofibers. In: Applications of Polymer Nanofibers. 2022. pp. 372-403

[7] Schnitzer O, Yariv E. The Taylor–Melcher leaky dielectric model as a macroscale electrokinetic description. Journal of Fluid Mechanics. 2015;**773**:1-33

[8] Luo C, Stoyanov SD, Stride E, Pelan E, Edirisinghe M. Electrospinning versus fibre production methods: From specifics to technological convergence. Chemical Society Reviews. 2012;**41**:4708-4735

[9] Tucker N, Stanger JJ, Staiger MP, Razzaq H, Hofman K. The history of the science and technology of electrospinning from 1600 to 1995. Journal of Engineered Fibers and Fabrics. 2012;7:63-73

[10] Bhagure SS, Rao AR. A review: Electrospinning and electrospinning nanofiber technology, process & application. International Journal of Innovative Science and Research Technology. 2020;**5**:528-538

[11] Zhao Y, Xiong J. Advanced scaffold design via electrospinning. In: Biomaterial Based Approaches to Study the Tumour Microenvironment. London: The Royal Society of Chemistry; 2022. pp. 197-225

[12] Li Y, Zhu J, Cheng H, Li G, Cho H, Jiang M, et al. Developments of advanced electrospinning techniques: A critical review. Advanced Materials Technologies. 2021;**6**:2100410

[13] Islam MS, Rana S, Islam MR, Faisal M, Rahman H, Sultana J. Porous core photonic crystal fibre for ultra-low material loss in THz regime. IET Communications. 2016;**10**:2179-2183

[14] Elahi MF, Lu W, Guoping G, Khan F. Core-shell fibers for biomedical applications-a review. Journal of Bioengineering and Biomedical Sciences. 2013;**3**:1-14

[15] Zhou X, Fang S, Leng X, Liu Z, Baughman RH. The power of fiber twist. Accounts of Chemical Research. 2021;**54**:2624-2636

[16] Bakeri G, Ismail AF, Shariaty-Niassar M, Matsuura T. Effect of polymer concentration on the structure and performance of polyetherimide hollow

fiber membranes. Journal of Membrane Science. 2010;**363**:103-111

[17] Wu Q-L, Zhao Y, Zhang Y-N, Yang Y. Characteristics of a new multi-channel sensing device based on C-type photonic crystal fibers. Optics & Laser Technology. 2021;**134**:106622

[18] Puppi D, Chiellini F. Wet-spinning of biomedical polymers: From single-fibre production to additive manufacturing of three-dimensional scaffolds. Polymer International. 2017;**66**:1690-1696

[19] Edie D. The effect of processing on the structure and properties of carbon fibers. Carbon. 1998;**36**:345-362

[20] Tkatch VI, Limanovskii AI, Denisenko SN, Rassolov SG. The effect of the melt-spinning processing parameters on the rate of cooling. Materials Science and Engineering: A. 2002;**323**:91-96

[21] Lovett ML, Cannizzaro CM, Vunjak-Novakovic G, Kaplan DL. Gel spinning of silk tubes for tissue engineering. Biomaterials. 2008;**29**:4650-4657

[22] Khan WS, Asmatulu R, Ceylan M, Jabbarnia A. Recent progress on conventional and non-conventional electrospinning processes. Fibers and Polymers. 2013;**14**:1235-1247

[23] Yoon J, Yang HS, Lee BS, Yu WR. Recent progress in coaxial electrospinning: New parameters, various structures, and wide applications. Advanced Materials. 2018;**30**:1704765

[24] Niu H, Wang X, Lin T. Needleless electrospinning: Developments and performances. In: Nanofibers-Production, Properties and Functional Applications. London, UK: IntechOpen; 2011. pp. 17-36

[25] Zhang C, Feng F, Zhang H. Emulsion electrospinning: Fundamentals, food applications and prospects. Trends in Food Science & Technology. 2018;**80**:175-186

[26] Badrossamay MR, McIlwee HA, Goss JA, Parker KK. Nanofiber assembly by rotary jet-spinning. Nano Letters. 2010;**10**:2257-2261

[27] He J-H, Kong H-Y, Yang R-R, Dou H, Faraz N, Wang L, et al. Review on fiber morphology obtained by bubble electrospinning and blown bubble spinning. Thermal Science. 2012;**16**:1263-1279

[28] Muerza-Cascante ML, Haylock D, Hutmacher DW, Dalton PD. Melt electrospinning and its technologization in tissue engineering. Tissue Engineering Part B: Reviews. 2015;**21**:187-202

[29] Lv H, Yu D-G, Wang M, Ning T. Nanofabrication of janus fibers through side-by-side electrospinning-a mini review. Materials Highlights. 2021;**2**:18-22

[30] Varesano A, Carletto RA, Mazzuchetti G. Experimental investigations on the multi-jet electrospinning process. Journal of Materials Processing Technology. 2009;**209**:5178-5185

[31] Shi X, Zhou W, Ma D, Ma Q, Bridges D, Ma Y, et al. Electrospinning of nanofibers and their applications for energy devices. Journal of Nanomaterials. 2015;**16**:122-122

[32] Syam Sundar L, Mesfin S, Tefera Sintie Y, Punnaiah V, Chamkha AJ, Sousa A. A review on the use of hybrid nanofluid in a solar flat plate and parabolic trough collectors and its enhanced collector thermal efficiency. Journal of Nanofluids. 2021;**10**:147-171

[33] Aadil M, Hassan W, Somaily HH, Ejaz SR, Abass RR, Jasem H, et al. Synergistic effect of doping and nanotechnology to fabricate highly efficient photocatalyst for environmental remediation. Journal of Alloys and Compounds. 2022;**920**:165876

[34] Zhang X, Xie L, Wang X, Shao Z, Kong B. Electrospinning super-assembly of ultrathin fibers from single-to multi-Taylor cone sites. Applied Materials Today. 2022;**26**:101272

[35] Stanger J, Tucker N, Kirwan K, Staiger MP. Effect of charge density on the Taylor cone in electrospinning. International Journal of Modern Physics B. 2009;**23**:1956-1961

[36] Bhardwaj N, Kundu SC. Electrospinning: A fascinating fiber fabrication technique. Biotechnology Advances. 2010;**28**:325-347

[37] Strawhecker KE, Kumar SK, Douglas JF, Karim A. The critical role of solvent evaporation on the roughness of spin-cast polymer films. Macromolecules. 2001;**34**:4669-4672

[38] Theron S, Zussman E, Yarin A. Experimental investigation of the governing parameters in the electrospinning of polymer solutions. Polymer. 2004;**45**:2017-2030

[39] Fang J, Wang H, Niu H, Lin T, Wang X. Evolution of fiber morphology during electrospinning. Journal of Applied Polymer Science. 2010;**118**:2553-2561

[40] Casper CL, Stephens JS, Tassi NG, Chase DB, Rabolt JF. Controlling surface morphology of electrospun polystyrene fibers: Effect of humidity and molecular weight in the electrospinning process. Macromolecules. 2004;**37**:573-578

Chapter 2

Physics of Electrospinning

Sharvare Palwai

Abstract

Electrospinning is the process of producing fibers ranging from sub-micron to Nano-scale in diameter consistently and reproducibly. The Electrospinning consists of three main parts High voltage power source (up to 30 kV), Spinneret (such as a syringe, with a small diameter needle) and a conducting collector. The basic principle of electrospinning technique is that, when an electrically charged solution is feed through a small opening such as syringe pump, needle or a pipette tip then due to its charge the solution is drawn as a jet towards an oppositely charged conducting collector plate. The solvent evaporates gradually during jet travel towards the collecting plate and a charged solid fiber is laid to accumulate at the collector plate. The high voltage is connected to the end of a needle containing the liquid solution. The fiber collecting screen is expected to be conductive and it can either be a stationary plate or a rotating platform or substrate. The physics of electrospinning involves several key factors, including the electrostatic forces, surface tension and viscosity of the polymer solution.

Keywords: nanofibers, membranes, spinneret, sensor, nanoscale polymers

1. Introduction

Electrospinning is a versatile and widely used technique for producing nanofibers with unique morphologies and properties. It has been studied and developed for over a century, with the first reported observation of electrospinning dating back to the late 1800s. Over the years, electrospinning has undergone several advancements, including the development of new equipment and methods to control the morphology and properties of the resulting nanofibers [1]. In this introduction, we will briefly discuss the history of the electrospinning process and some notable milestones and achievements that have contributed to its development.

The earliest known observation of electrospinning was made by British scientist William Gilbert in the late 1600s. However, it was not until the late 1800s that electrospinning was first reported as a process for producing fibers. In 1884, French scientist Charles Vincent reported the production of fibers by electrostatic means. The first patent for electrospinning was granted in 1900 to American scientist Joseph Zeleny, who used the technique to produce fibers for medical applications. In the 1930s and 1940s, electrospinning gained attention as a method for producing ultrafine fibers for air filtration and other applications. In the 1960s, the use of electrospinning expanded to include the production of synthetic fibers for textile applications. In the 1990s and

2000s, electrospinning became a popular research tool for the production of nanofibers for a wide range of applications, including tissue engineering, drug delivery, and energy storage. Since then, electrospinning has undergone significant advancements and innovations. Researchers have developed new equipment and methods to control the morphology and properties of the resulting nanofibers, including the use of multi-needle spinnerets, coaxial and triaxial electrospinning, and the addition of other materials to the polymer solution [2]. These advancements have expanded the potential applications of electrospinning and opened up new avenues for research.

Fiber production techniques have been the focus of much research over the years, with electrospinning being one of the most promising due to its versatility, adaptability, and the unique properties of the fibers produced [3]. However, other traditional and contemporary fiber formation techniques, such as melt spinning, dry spinning, wet spinning, and centrifugal spinning, still hold substantial relevance in various applications [4]. This review explores the advancements and the comparative strengths and weaknesses of electrospinning versus other fiber production techniques. Electrospinning stands out as a simple and versatile technique for creating ultrafine fibers with diameters ranging from the Nano to the microscale [5]. It offers exceptional control over fiber morphology and structure, enabling the production of complex nanofiber arrangements [3]. Its versatility in handling a wide array of materials, such as polymers, ceramics, and composites, is noteworthy [6]. Furthermore, electrospun fibers exhibit unique properties, such as high surface area-to-volume ratio, flexibility in surface functionalities, and superior mechanical properties, which are attractive for applications in filtration, tissue engineering, drug delivery, and energy storage [7]. Despite these strengths, electrospinning faces a major limitation in its low production rate, rendering it less feasible for large-scale industrial applications [3]. Some techniques have attempted to overcome this limitation, such as needleless electrospinning and multi-jet electrospinning, but the scalability issue remains a challenge [8]. On the other hand, traditional fiber spinning techniques, like melt, dry, and wet spinning, have been utilized for large-scale fiber production for years [4]. These methods enable the production of continuous fibers and provide significant control over the fiber diameter, although generally at a larger scale than electrospinning. However, these techniques struggle to match the nanoscale dimensions and high surface area that electrospinning can achieve. Additionally, the processing conditions in traditional spinning methods, such as high temperature in melt spinning and solvent use in dry and wet spinning, can limit the materials that can be used [9]. Centrifugal spinning, a more recent development, emerges as an alternative to electrospinning, offering higher throughput and the ability to produce fibers in the Nano to microscale range [10]. It eliminates the use of high voltages, making it safer, and offers more flexibility in the types of materials that can be spun [11]. Yet, the fibers produced by this method still lack the uniformity achieved by electrospinning, and the method requires significant optimization to improve the fiber properties and morphology [10]. In conclusion, electrospinning continues to hold a distinctive position among fiber production techniques due to its capability of generating unique fiber properties. Despite the competition from traditional and newer techniques, the challenges of scale-up production in electrospinning highlight the importance of ongoing research in enhancing this technology's feasibility for industrial applications.

In the field of biomedical applications, electrospun fibers have shown immense promise, particularly in tissue engineering, wound healing, and drug delivery systems. They exhibit a structural similarity to the natural extracellular matrix, which promotes cellular growth and tissue regeneration ("Electrospun nanofibers for regenerative medicine," 2020, Advanced Drug Delivery Reviews). In the realm of energy storage, electrospun nanofibers have emerged as compelling candidates for creating high-performance electrodes in lithium-ion batteries and super capacitors. Their high surface-to-volume ratio facilitates better electron transfer and ion diffusion, contributing to improved energy storage efficiency ("Nanofibers for lithium-ion and lithium-metal batteries: A Review," 2020, Advanced Materials). Environmental remediation represents another significant domain of electrospun fiber applications. Electrospun nanofibers, especially those functionalized with certain nanomaterials or specific chemical groups, have demonstrated excellent capabilities for absorbing heavy metals, organic pollutants, and oil spills from water ("Electrospun nanofibers for environmental applications," 2019, Nanomaterials). The filtration industry has also harnessed the potential of electrospun fibers, thanks to their adjustable pore sizes and high porosity. They have been proven effective for filtering out particulate matter, bacteria, and viruses from air and liquids ("Filtration properties of electrospun Nano fibrous materials for aeronautic applications," 2015, Journal of Aerosol Science). Lastly, electrospun fibers' high sensitivity and specific surface area make them excellent for sensor technologies, where they can be functionalized to detect various chemicals, gases, and biomaterials ("Electrospun nanofibers for chemical and biological sensors," 2020, Materials Today Chemistry) [3, 12–19].

In conclusion, the electrospinning process has a long and rich history, with significant advancements and innovations made over the years. It has become a widely used technique for producing nanofibers with unique morphologies and properties, with potential applications in a wide range of fields. The continued development and improvement of the electrospinning process hold great promise for the future of nanofiber research and applications [20].

2. Experimental techniques

The electrospinning setup consists of several key components, including a high-voltage power supply, a spinneret, a collector, and a syringe pump. While the basic electrospinning setup is relatively simple, advancements in the field have led to the development of more advanced and unique setups that offer greater control over the morphology and properties of the resulting nanofibers. Following are some of the set-up's.

2.1 Single needle setup

One of the earliest and simplest electrospinning setups involves the use of a single-needle spinneret as shown in **Figure 1** below. This setup involves the attachment of a syringe to a metallic needle, which is then connected to a high-voltage power supply. The polymer solution or melt is loaded into the syringe, which is then driven by a syringe pump to control the flow rate. The polymer solution is then ejected through the needle and forms a charged jet as it travels towards a collector plate. Researchers from the School of Chemical Engineering at the University of Adelaide, Australia used a single-needle spinneret electrospinning setup to fabricate poly(lactic acid) (PLA) nanofibers for biomedical applications. In their study, they loaded a PLA solution into a syringe attached to a 22-gauge metallic needle and connected it to a high-voltage power supply. The solution was then ejected through the needle and formed a charged jet that was collected on a

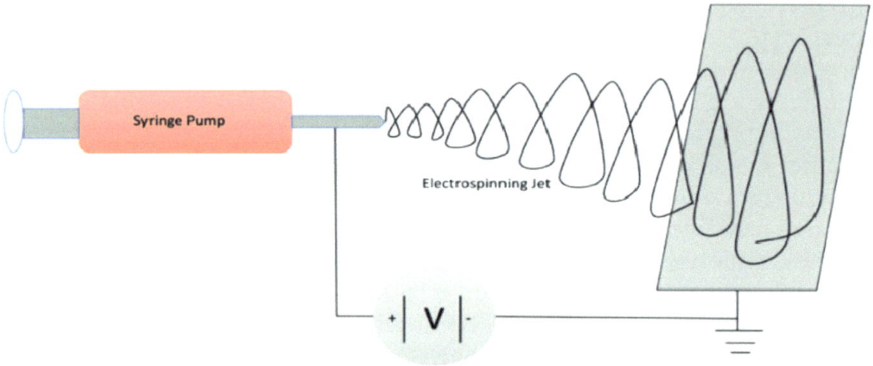

Figure 1.
Basic electrospinning set-up with a single needle [21].

grounded aluminum foil collector. The researchers varied the flow rate of the PLA solution using a syringe pump to control the diameter of the resulting nanofibers. They observed that the nanofibers exhibited high surface area-to-volume ratios, which could be beneficial for drug delivery applications [22].

2.2 Multi needles setup

Another commonly used setup is the multi-needle electrospinning setup as shown in **Figure 2**. This setup involves the use of multiple needles arranged in a specific pattern to electrospun several fibers simultaneously. Multi-needle electrospinning is useful for the production of large quantities of nanofibers in a short amount of time and for the production of more complex fiber morphologies. Custom-built electrospinning setup that consisted of six needles arranged in a circular pattern were used. The needles were connected to a high-voltage power

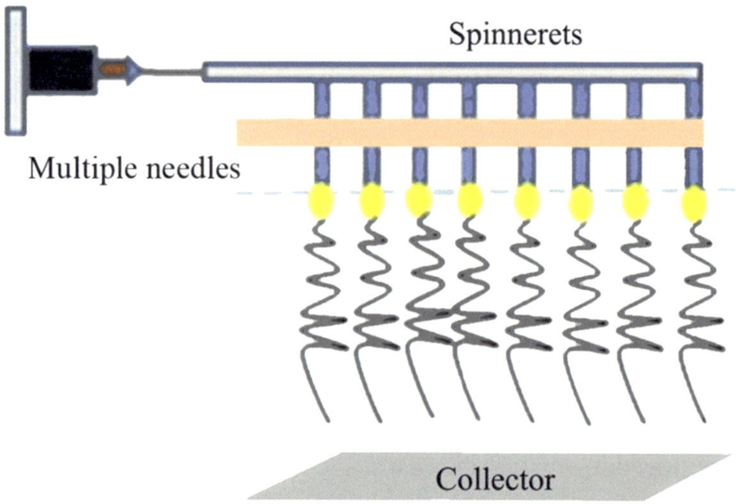

Figure 2.
Multi-needle electrospinning equipment [23].

supply and a syringe pump was used to control the flow rate of the polycaprolactone (PCL) solution. The solution was then electrospun into nanofibers, which were collected on a rotating cylindrical drum. The researchers varied the distance between the needles and the collector to control the alignment of the nanofibers. They observed that the aligned PCL nanofibers had enhanced mechanical properties and could promote cell adhesion and proliferation [24].

2.3 Needle-less electrospinning setup

Needleless electrospinning is a technique used to produce nanofibers from a polymer solution or melt without using a needle as shown in **Figure 3**. Instead, the polymer solution is extruded from a capillary orifice, which is usually a small nozzle or spinneret. The process is similar to traditional electrospinning, but the use of a capillary orifice instead of a needle has some advantages [26, 27].

One advantage of needleless electrospinning is that it reduces the risk of contamination. In traditional electrospinning, the needle can become contaminated with the polymer solution, which can affect the quality of the nanofibers produced. With needleless electrospinning, the risk of contamination is reduced because there is no needle to become contaminated.

Another advantage of needleless electrospinning is that it can produce fibers with a more uniform size distribution as shown in **Figure 4**. This is because the capillary orifice can be designed to produce a more uniform flow of polymer solution, which can result in more uniform nanofibers. In contrast, traditional electrospinning can produce nanofibers with a wider size distribution due to variations in the flow rate of the polymer solution from the needle [29].

Needleless electrospinning can be achieved using various techniques, such as centrifugal electrospinning, electro spraying, and electro hydrodynamic atomization as shown in **Figure 5**. Each technique has its advantages and disadvantages, and the selection of the appropriate technique depends on the specific application and requirements of the nanofiber product [5, 31].

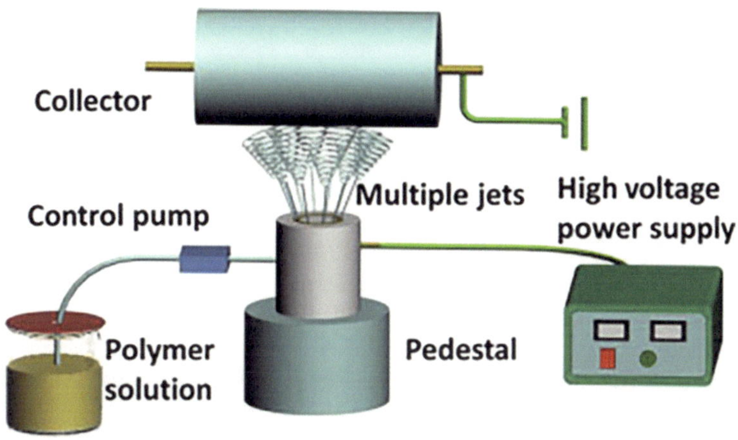

Figure 3.
Needleless electrospinning [25].

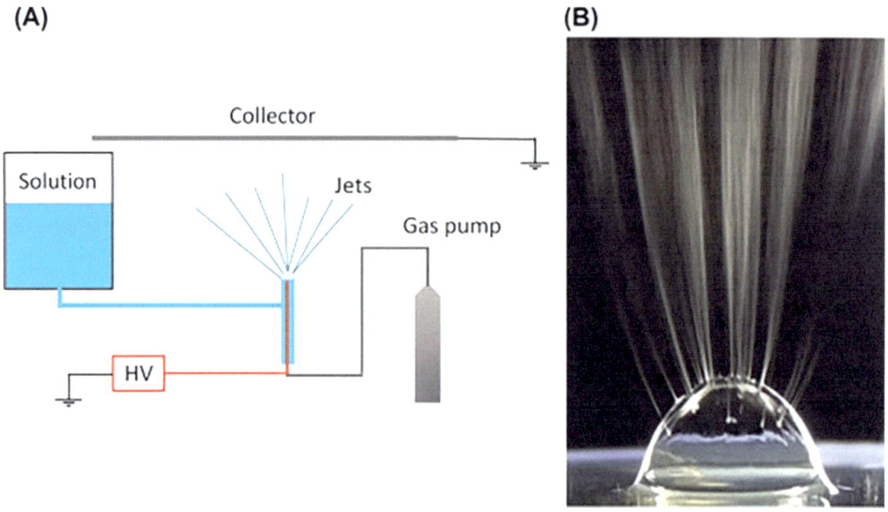

Figure 4.
Needle-less electro spraying [28].

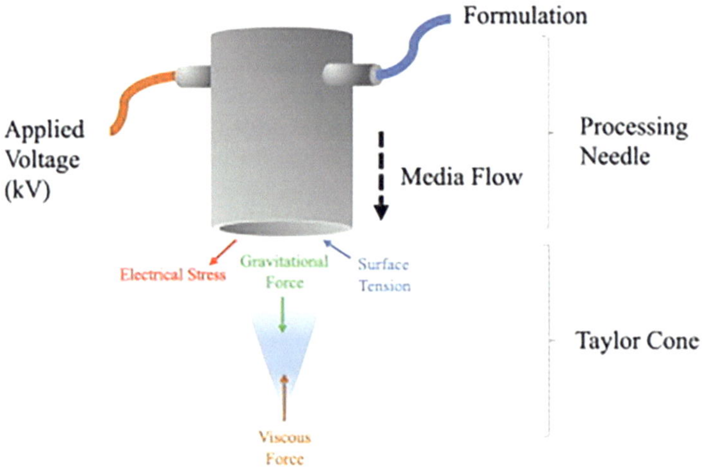

Figure 5.
Needleless electro hydrodynamic atomization [30].

2.4 Coaxial electrospinning setup

This is another advanced electrospinning setup used to produce core-shell nanofibers. This setup involves the use of a coaxial spinneret, which consists of two concentric tubes. The polymer solution or melt is loaded into the inner tube, and a different polymer solution or melt is loaded into the outer tube. The polymer solution from the inner tube forms the core of the nanofiber, while the polymer solution from the outer tube forms the shell as shown in **Figure 6**. For example, researchers have used coaxial electrospinning to produce nanofibers with a drug-loaded core and a polymer shell for controlled drug delivery applications [33]. In the study conducted by [34] they loaded a poly lactic-co-glycolic acid (PLGA) solution

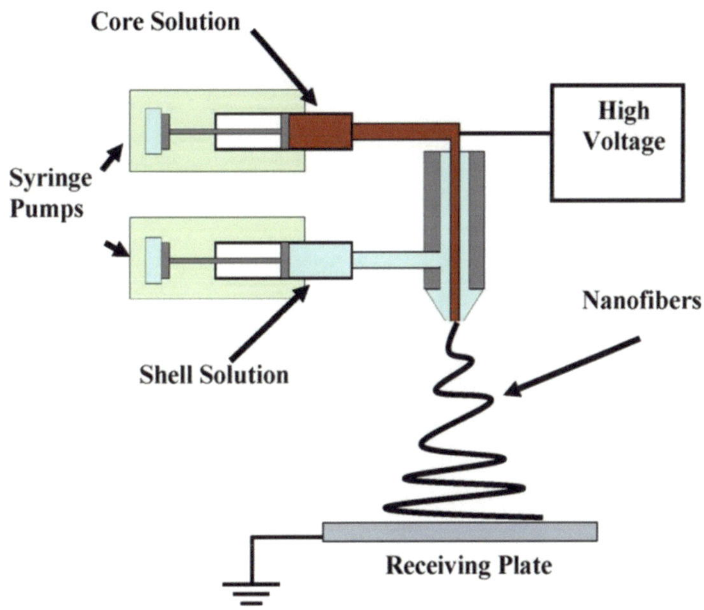

Figure 6.
Co-axial electrospinning equipment [32].

containing a model drug into the inner tube of the coaxial spinneret and a PLGA solution without the drug into the outer tube. The two solutions were electrospun simultaneously to form core-shell nanofibers, which were collected on a grounded aluminum foil. The researchers varied the flow rate and composition of the two solutions to control the drug loading and release profile of the resulting nanofibers. They observed that the nanofibers exhibited sustained drug release over a period of several weeks, demonstrating the potential of this technique for controlled drug delivery applications.

2.5 Microfluidic electrospinning setup

This setup is used for the production of nanofibers with precise control over their morphology and properties. This setup involves the use of microfluidic channels to control the flow rate and concentration of the polymer solution before it is electrospun as shown in **Figure 7**. Microfluidic electrospinning is useful for the production of nanofibers with more complex morphologies and structures, such as patterned and multi-layered fibers. Researchers have used microfluidic electrospinning to produce nanofibers with a gradient of pore size for tissue engineering applications [36]. In the study conducted by [37] they used a microfluidic chip consisting of two parallel channels, one for the PCL solution and the other for a pore-forming agent. The two solutions were then electrospun through a single needle to produce nanofibers with a gradient of pore size. The nanofibers were collected on a grounded aluminum foil and then treated with ethanol to remove the pore-forming agent. The researchers observed that the resulting nanofibers exhibited a gradient of pore size, which could be beneficial for tissue engineering applications.

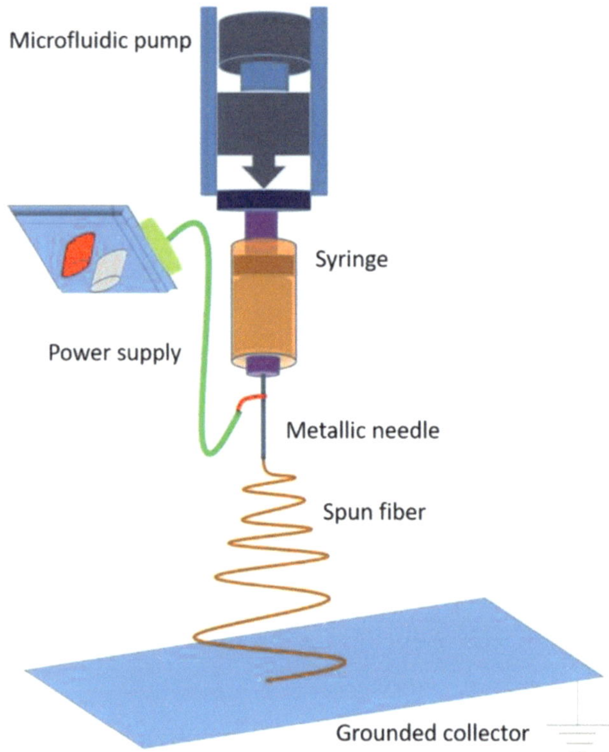

Figure 7.
Microfluidic pump electrospinning setup [35].

3. Electrospinning process

The physics of electrospinning is complex and involves several mechanisms, including electrostatic repulsion, surface tension, and solvent evaporation. The process begins when a high voltage is applied to a droplet of polymer solution, creating an electric field that causes charges to accumulate on the surface of the droplet. When the electric field reaches a critical value, electrostatic repulsion overcomes surface tension and a thin jet of material is ejected from the droplet. As the jet travels through the air, solvent evaporation causes it to stretch and thin, resulting in the formation of a fiber.

Several factors can influence the electrospinning process, including the properties of the polymer solution (such as viscosity, conductivity, and surface tension), the applied voltage, the distance between the spinneret and the collector, and the properties of the collector (such as its shape and surface energy). Understanding the physics behind electrospinning is important for optimizing the process and producing fibers with desired properties. The process can be divided into three main stages: (1) initiation, (2) elongation, and (3) solidification as shown in **Figure 8**.

3.1 Initiation phase

In the initiation stage, a high voltage is applied to the polymer solution or melt. The voltage creates an electric field that induces charges on the surface of the solution, resulting in the formation of a Taylor cone. The Taylor cone is a conical shape

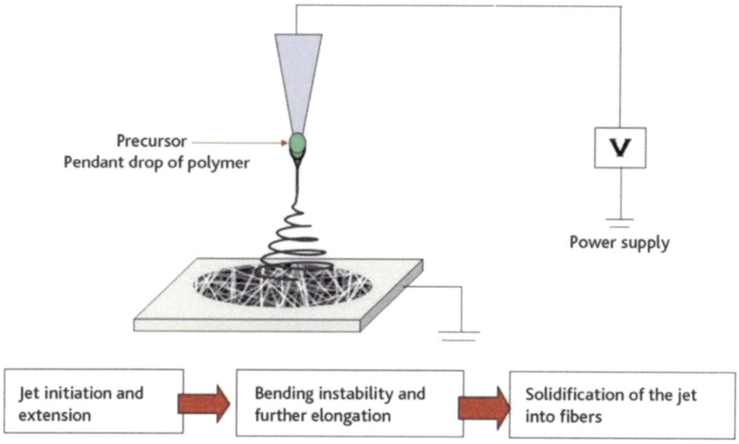

Figure 8.
Basic Schematic of three main phases in electrospinning technique [38].

formed at the tip of the spinneret, where the electric field strength is the highest. The formation of the Taylor cone is a critical step in the electrospinning process, as it determines the initial shape of the jet. The formation of the jet involves a complex interplay of electrical, fluidic, and surface tension forces, which are influenced by several factors such as the properties of the polymer solution, the electric field strength, and the distance between the spinneret and the collector.

The electrostatic force is the driving force behind the initiation of the jet. When a high voltage is applied to a conductive polymer solution or melt, charges accumulate on the surface of the solution. This accumulation of charges creates an electrostatic field that induces a polarization of the solution. As the electric field strength increases, the polarization becomes more pronounced, and the surface tension of the solution decreases [39]. Eventually, the surface tension becomes so low that it can no longer support the weight of the solution, and a droplet forms at the tip of the spinneret.

Once the droplet forms, the electrical repulsion between the charges on the surface of the droplet overcomes the surface tension, and a charged jet is ejected from the droplet. The charged jet then experiences a series of instabilities, such as the Rayleigh instability, which cause the jet to break up into droplets. The droplets solidify into nanofibers as they travel towards the collector [40].

Polymer solution properties: The viscosity of the polymer solution affects the formation of the droplet at the tip of the spinneret, while the conductivity and surface tension influence the formation and stability of the charged jet. A high conductivity can result in a more stable jet, while a low surface tension can promote the formation of a more uniform jet.

Electric field strength: A higher electric field strength can lead to the formation of a more stable and uniform jet, but can also result in the formation of multiple jets or the formation of beads along the fiber. The electric field strength can be controlled by adjusting the voltage applied to the spinneret and the collector [2].

Distance between the spinneret and the collector: The distance can influence the size and shape of the droplet and the jet, as well as the alignment and morphology of the resulting nanofibers. A shorter distance can result in a more uniform jet and more aligned fibers, while a longer distance can result in a more random orientation of the fibers [41, 42].

Ambient conditions: High humidity can increase the surface tension of the polymer solution and lead to the formation of larger droplets and less stable jets.

3.2 Elongation phase

In the elongation stage, the electric field causes the surface of the Taylor cone to become unstable, resulting in the formation of a jet. The jet is propelled towards the collector due to the repulsive electrostatic forces between the charges on the surface of the jet and the charges on the collector. As the jet travels through the air, it undergoes elongation and thinning due to a combination of electrostatic repulsion, surface tension, and air resistance. The elongation stage of electrospinning is a critical step that determines the final morphology and properties of the resulting nanofibers.

The viscosity of the polymer solution can be controlled by adjusting the concentration of the polymer or by adding a viscosity modifier, such as a polymer solvent or a surfactant. The conductivity of the polymer solution can be controlled by adding conductive additives, such as carbon nanotubes or graphene. The surface tension of the polymer solution can be controlled by adding a surfactant or by adjusting the pH of the solution. A study by Huang et al. [39] investigated the effect of polymer concentration on the morphology and properties of electrospun nanofibers. The study found that increasing the polymer concentration resulted in thicker fibers with larger diameters, while decreasing the concentration led to thinner fibers with smaller diameters. This highlights the importance of controlling the viscosity of the polymer solution during the elongation stage of electrospinning.

Another approach to controlling the elongation stage of electrospinning is to modify the electric field strength. This can be achieved by adjusting the voltage applied to the spinneret and the collector or by using a multi-needle spinneret to generate multiple jets with different electric field strengths. A higher electric field strength can result in a faster stretching rate and thinner fibers, while a lower electric field strength can result in thicker fibers. A study by Zeng et al. [43] explored the use of multi-needle electrospinning to control the morphology and properties of electrospun nanofibers. The study found that using a multi-needle spinneret resulted in the formation of multiple jets with different electric field strengths, which allowed for greater control over the fiber diameter and morphology. This highlights the potential of using advanced electrospinning techniques to improve the precision and reproducibility of electrospun nanofibers.

A shorter distance can result in a faster stretching rate and thinner fibers, while a longer distance can result in thicker fibers. However, a shorter distance can also increase the likelihood of secondary droplet formation along the jet, which can result in the formation of beads along the fiber.

High humidity can increase the surface tension of the polymer solution, leading to the formation of thicker fibers. High temperature can decrease the viscosity of the polymer solution, resulting in faster stretching rates and thinner fibers. However, high temperature can also lead to faster solvent evaporation, affecting the final morphology and properties of the resulting nanofibers. In a study by Khajavi et al. [44] the effect of humidity on the morphology and properties of electrospun nanofibers was investigated. The study found that increasing humidity led to an increase in fiber diameter due to an increase in surface tension. This demonstrates the importance of controlling the environmental conditions during the elongation stage of electrospinning to achieve the desired fiber morphology and properties.

For example, the use of co-electrospinning, where multiple polymers are electrospun simultaneously, can be used to control the morphology and properties of the resulting nanofibers. Electrospinning with the aid of external fields, such as magnetic or acoustic fields, can also be used to control the elongation stage and orientation of the resulting nanofibers. In a study by Lee et al. [45] the effect of surfactant concentration on the morphology and properties of electrospun nanofibers was investigated. The study found that increasing the surfactant concentration led to a decrease in fiber diameter due to a decrease in surface tension. This demonstrates the potential of surfactants as a means of controlling the elongation stage of electrospinning.

3.3 Solidification phase

In the solidification stage, the solvent evaporates from the jet, resulting in the solidification of the polymer. The rate of solvent evaporation is critical in determining the final morphology of the fibers. Rapid solvent evaporation leads to the formation of smooth fibers, while slow solvent evaporation leads to the formation of beaded fibers. In a study by Scaffaro et al. [46] the effect of solvent evaporation rate on the morphology and properties of electrospun nanofibers was investigated. The study found that a rapid solvent evaporation rate led to the formation of smooth fibers, while a slow evaporation rate led to the formation of beaded fibers. This highlights the importance of controlling the rate of solvent evaporation during the solidification stage of electrospinning.

High viscosity and surface tension lead to the formation of thicker fibers, while low viscosity and surface tension lead to the formation of thinner fibers. Conductive solutions promote the formation of straight fibers due to the higher electrostatic forces acting on the charged jet. Increasing the concentration of the polymer solution can lead to the formation of thicker fibers, but can also lead to the formation of beaded fibers due to the slower solvent evaporation.

Higher voltages lead to the formation of thinner fibers due to the increased electrostatic forces acting on the charged jet. However, excessively high voltages can lead to the formation of irregular and discontinuous fibers. In a study by Reneker et al. [47] the effect of applied voltage on the morphology and properties of electrospun nanofibers was investigated. The study found that higher voltages led to the formation of thinner fibers due to the increased electrostatic forces acting on the charged jet. However, excessively high voltages can lead to the formation of irregular and discontinuous fibers. This highlights the importance of optimizing the applied voltage to achieve the desired fiber morphology and properties.

Shorter distances lead to the formation of thinner fibers due to the reduced air resistance acting on the charged jet. A study by Lee et al. [48] investigated the effect of distance between the spinneret and collector on the morphology and properties of electrospun nanofibers. The study found that shorter distances led to the formation of thinner fibers due to the reduced air resistance acting on the charged jet. This demonstrates the potential of controlling the spinneret-collector distance to achieve the desired fiber morphology during the electrospinning process.

4. Conclusion

The field of electrospinning is a captivating amalgamation of physics and material science. As a technological process, it has carved out a unique position, enabling the generation of micro and nanoscale fibers with diverse applications across industries such as biomedical, textiles, energy, and environmental science. Electrospinning relies on the principles of electro hydrodynamics, exploiting the Columbic forces to create a high-speed jet of polymer solution or melt. As the solvent evaporates, it leaves behind a thin fiber that can be collected in various forms, depending on the configuration of the collector. The high surface area to volume ratio, direct result of the physics of the process, enables electrospun materials to have enhanced properties in terms of reactivity, sensitivity, and diffusivity, thus overcoming the limitations of traditional fiber fabrication methods. One of the critical advantages of electrospinning is the versatility of fiber diameter it can produce, ranging from a few nanometers to several micrometers, by merely adjusting the solution properties and process parameters. The ability to generate ultrafine fibers consistently grants this technique an edge over conventional methods like drawing or phase separation, whose outcome can vary significantly. Moreover, electrospinning offers the flexibility to create fibers with tailored properties, including the porosity, surface chemistry, and orientation, thus opening up numerous possibilities for applications where these properties play a pivotal role. From bioengineering, where the structure of electrospun nanofibers can mimic the extracellular matrix for tissue scaffolding, to filtration and protective clothing applications, where pore size control is crucial, the ability to customize these parameters offers a distinct advantage. The simplicity and versatility of the setup not only provide a cost-effective solution but also allow for various adaptations and modifications. For instance, using multiple needles or a free surface setup for co-axial or multilayer fibers, implementing a rotating or patterned collector for aligned or patterned fiber mats, or adding post-processing steps for further fiber modifications. In conclusion, the physics of electrospinning has revolutionized the fabrication of micro and nanofibers, offering unprecedented advantages over traditional methods. Its ability to generate ultrafine fibers with a high surface area to volume ratio, the versatility in fiber properties, and a straightforward, adaptable setup make electrospinning a powerful technique with broad-spectrum applications. As we delve deeper into the realm of nanotechnology and advanced materials, electrospinning continues to hold enormous potential for novel, high-impact solutions across various fields of science and technology.

Author details

Sharvare Palwai
Physics, Tuskegee University, United States

*Address all correspondence to: psharvare@gmail.com

References

[1] Lu P, Yu H. Handbook of Electrospinning. Boca Raton, Florida, USA: CRC Press; 2008

[2] Reneker DH, Chun I. Nanometer diameter fibers of polymer, produced by electrospinning. Nanotechnology. 1996;**7**(3):216

[3] Zhang S et al. Recent developments in electrospinning design for advanced nanofibers: A review. Journal of Materials Science & Technology. 2022;**101**:46-65

[4] Li M et al. Advances in high-throughput production of micro- and nano-fibers: From traditional techniques to recent developments. Journal of Applied Polymer Science. 2023;**140**(3):487-503

[5] Sill TJ, von Recum HA. Electrospinning: Applications in drug delivery and tissue engineering. Biomaterials. 2008;**29**:1989-2006. DOI: 10.1016/j.biomaterials.2008.01.011

[6] Liu J, He J. A comprehensive review of electrospinning for tissue engineering applications. Advanced Healthcare Materials. 2022;**11**(4):e2101016

[7] Reneker DH, Chun I. Nanometer diameter fibers of polymer, produced by electrospinning. Nanotechnology. 1996;**7**(3):216-223

[8] Park J et al. Needleless electrospinning for scalable production of nanofibers: Progress, challenges, and opportunities. Materials Today. 2023;**36**:100-120

[9] Jiang Y et al. Polymer fibers: Melt, wet, and dry spinning. Polymer Science. 2021;**63**(1):45-57

[10] Elshereef R et al. Comparative study of nanofibers produced by centrifugal spinning and electrospinning. Journal of Materials Science. 2022;**57**(10):6734-6749

[11] Liu Y et al. Centrifugal spinning: A review of material systems and future applications. Journal of Engineered Fibers and Fabrics. 2023;**18**(1):15589250220963890

[12] Sill TJ, von Recum HA. Electrospinning: Applications in drug delivery and tissue engineering. Biomaterials. 2021;**32**(22):5054-5060

[13] Bhardwaj N, Kundu SC. Electrospinning: A fascinating fiber fabrication technique. Biotechnology Advances. 2010;**28**(3):325-347

[14] Ding B, Kim HY, Lee SC, Shao CL, Lee DR, Park SJ, et al. Preparation and characterization of a nanoscale poly(vinyl alcohol) fiber aggregate produced by an electrospinning method. Journal of Polymer Science Part B: Polymer Physics. 2006;**44**(1):246-253

[15] Yang D, Lu B, Zhao Y, Jiang X. Fabrication of aligned fibrous arrays by magnetic electrospinning. Advanced Materials. 2007;**19**(20):3702-3706

[16] Xue J, Wu T, Dai Y, Xia Y. Electrospinning and electrospun nanofibers: Methods, materials, and applications. Chemical Reviews. 2019;**119**(8):5298-5415

[17] Subbiah T, Bhat GS, Tock RW, Parameswaran S, Ramkumar SS. Electrospinning of nanofibers. Journal of Applied Polymer Science. 2005;**96**(2):557-569

[18] Reneker DH, Chun I. Nanometre diameter fibres of polymer, produced by electrospinning. Nanotechnology. 1996;7(3):216

[19] Xie J, Li X, Xia Y. Putting electrospun nanofibers to work for biomedical research. Macromolecular Rapid Communications. 2008;**29**(22):1775-1792

[20] Zhang YZ, Venugopal J, Huang ZM, Lim CT, Ramakrishna S. Characterization of the surface biocompatibility of the electrospun PCL-collagen nanofibers using fibroblasts. Biological Macromolecule. 2006;7(4):1049-1055

[21] Palwai S. Impact of perovskite materials in ferroelectric polymer. Nanotechnology & Applications. 2020;**4**(1):1-5

[22] Munir A, Zohora FT, Chen X. Electrospun poly (lactic acid) (PLA) nanofibers for biomedical applications: A review. Journal of Materials Research. 2019;**34**(4):636-647. DOI: 10.1557/jmr.2018.456

[23] Ding B, Wang X, Yu J, editors. Electrospinning: Nanofabrication and Applications. Amsterdam: Elsevier; 2019

[24] Verma RK, Singh AK, Singh V. Multi-needle electrospun polycaprolactone nanofibers for wound healing applications. Polymer Engineering & Science. 2021;**61**(8):2016-2025. DOI: 10.1002/pen.25584

[25] Wei L, Sun R, Liu C, Xiong J, Qin X. Mass production of nanofibers from needleless electrospinning by a novel annular spinneret. Materials & Design. 2019;**179**:107885

[26] Li D, Xia Y. Electrospinning of nanofibers: Reinventing the wheel. Advanced Materials. 2004;**16**:1151-1170. DOI: 10.1002/adma.200400719

[27] Wang X, Ding B, Li B. Needleless electrospinning of nanofibers with a conical nozzle. Journal of Nanomaterials. 2013;**2013**:732647. DOI: 10.1155/2013/732647

[28] Yan G, Niu H, Lin T. Needle-less Electrospinning. In: Electrospinning: Nanofabrication and Applications. Amsterdam, The Netherlands: Elsevier; 2019

[29] Sun Z, Zussman E, Yarin AL, Wendorff JH, Greiner A. Compound core-shell polymer nanofibers by co-electrospinning. Advanced Materials. 2003;**15**:1929-1932. DOI: 10.1002/adma.200305304

[30] Ali A, Zaman A, Sayed E, et al. Electrohydrodynamic atomisation driven design and engineering of opportunistic particulate systems for applications in drug delivery, therapeutics and pharmaceutics. Advanced Drug Delivery Reviews. 2021;**176**:113788

[31] Xu L, Xu C, Yu H, Zhang W, Liu Y, Wu J. Centrifugal electrospinning for nanofiber production. Journal of Nanomaterials. 2013;**2013**:717315. DOI: 10.1155/2013/717315

[32] Chen R, Huang C, Ke Q, He C, Wang H, Mo X. Preparation and characterization of coaxial electrospun thermoplastic polyurethane/collagen compound nanofibers for tissue engineering applications. Colloids and Surfaces B: Biointerfaces. 2010;**79**(2):315-325

[33] Xue J, Wu T, Dai Y, Xia Y, Yang W. Electrospinning and electrospun nanofibers: Methods, materials, and applications. Chemical Reviews. 2010;**119**(8):5296-5415

[34] Jia X, Wang H, Yan X, Zhu L, Chen W. Coaxial electrospinning for

encapsulation and controlled release of fragile water-soluble bioactive agents. Journal of Controlled Release. 2017;**259**:e80. DOI: 10.1016/j.jconrel.2017.02.029

[35] Mondal K. Recent advances in the synthesis of metal oxide nanofibers and their environmental remediation applications. Inventions. 2017;**2**(2):9

[36] Xu J, Li Y, Li X, Gu Z. A hybrid electrospinning-microfluidic method for the preparation of high-quality protein nanofibers for drug delivery. Colloids and Surfaces B: Bio Interfaces. 2019;**177**:55-62

[37] Huang J, Grunlan MA, Qin X. Microfluidic electrospinning of gradient PCL/chitosan scaffolds for neural tissue engineering. Journal of Materials Chemistry B. 2016;**4**(14):2396-2404. DOI: 10.1039/c6tb00061g

[38] Mirzaei A, Kumar V, Bonyani M. Conducting polymer nanofibers based sensors for organic and inorganic gaseous compounds. Asian Journal of Atmospheric Environment. 2020;**14**(2):85-104

[39] Huang Z-M, Zhang YZ, Kotaki M, Ramakrishna S. A review on polymer nanofibers by electrospinning and their applications in nanocomposites. Composites Science and Technology. 2003;**63**(15):2223-2253

[40] Li D, Xia Y, Whiteside GM. Electrospinning of polymeric and ceramic nanofibers as uniaxial aligned arrays. Nature. 1996;**383**(6601):146-149

[41] Deitzel JM, Kleinmeyer J, Harris D, Beck Tan NC. The effect of processing variables on the morphology of electrospun nanofibers and textiles. Polymer. 2001;**42**(1):261-272

[42] Fong H, Reneker DH. Electrospinning and stabilization of polymer nanofibers. Advanced Materials. 1999;**11**(12):2-5

[43] Zeng J, Xu X, Chen X, Liang Q, Bian X, Yang L, et al. Controllable multiple jets electrospinning for the preparation of aligned polymer nanofibers. Materials Letters. 2018;**214**:135-138. DOI: 10.1016/j.matlet.2017.12.091

[44] Khajavi R, Abbasipour M, Bahador A. Effect of humidity on electrospinning of PCL/CS blend nanofibers. International Journal of Polymeric Materials and Polymeric Biomaterials. 2019;**68**(10):593-599. DOI: 10.1080/00914037.2018.1476057

[45] Lee JY, Bashur CA, Goldstein AS. Effect of surfactant concentration on electrospun polycaprolactone fiber morphology and diameter. Journal of Biomaterials Science, Polymer Edition. 2019;**30**(12):1066-1080. DOI: 10.1080/09205063.2019.1630797

[46] Scaffaro R, Lopresti F, Botta L. Electrospinning of polymer nanofibers: An overview on the solidification stage. Materials Science. 2020;**109**:2120-2136

[47] Reneker DH, Yarin AL, Fong H, Koombhongse S. Bending instability of electrically charged liquid jets of polymer solutions in electrospinning. Journal of Applied Physics. 2000;**87**(9):4531-4547. DOI: 10.1063/1.372871

[48] Lee KY, Jeong L, Kang YO, Lee SJ. Endothelialization of modified ePTFE grafts: A comparison study between electrospun and solution-cast biodegradable polyurethane scaffolds. Journal of Biomedical Materials Research Part A. 2006;**77**(4):705-714. DOI: 10.1002/jbm.a.30696

Chapter 3

Electrospinning-Based Super Liquid-Repellent Membranes for Membrane Distillation: Theory, Fabrications, Applications, and Challenges

Xiaocheng Zhang, Yuan Liao and Abdul Ghani Razaqpur

Abstract

The potential of membrane distillation (MD) in seawater desalination and high-salt wastewater treatment makes it a highly promising application in alleviating the global water crisis. However, membrane fouling and wetting are the main obstacles to the large-scale application of MD. Bio-inspired super liquid-repellent membranes offer a viable resolution to these challenges. The rapid advancement of nanotechnology has stimulated the growing attention toward electrospun nanofiber membranes (ENMs). Electrospun fibers demonstrate excellent functionalization, controllability, and hydrophobicity. Their low energy consumption and ease of preparation promote their application prospects in the construction of super liquid-repellent membranes. This article provides a comprehensive summary of electrospinning principles and influencing factors, coupled with a detailed account of the theory and preparation of super-liquid-repellent membranes *via* electrospinning, thus explicating the application and challenges of these membranes in MD, facilitating a deeper understanding of the ENMs application in MD for readers.

Keywords: membrane distillation, super liquid-repellent membranes, electrospinning, fouling, wetting

1. Introduction

Water is an essential resource for the survival of human society. However, only 2.5% of the earth's total water is fresh water, and most of it is deep groundwater, glaciers, and polar ice caps, rendering it unusable for humans [1]. As a result, less than 1% of the world's fresh water resources are available for human consumption and production. With the rapid growth of industrialization and population, the tension between water supply and demand has intensified, leading to an increasingly severe water scarcity problem. This has resulted in an estimated one billion people worldwide lacking access to clean fresh water. Therefore, finding ways to alleviate the shortage of fresh water resources has become a critical issue of global concern.

Desalination can facilitate the production of freshwater and alleviate the crisis of freshwater scarcity. Its two primary existing practices include thermal treatment and membrane technology [2]. Nevertheless, thermally driven processes for seawater desalination have been observed to be susceptible to scaling, which entail high energy consumption and expenses. As a result, the membrane technology has gained attraction in the market with its associated benefits [3]. Thanks to its low energy consumption and small space requirements, reverse osmosis (RO) is currently the most widely used desalination technology worldwide. However, a significant amount of concentrated salt wastewater is produced in the RO process, adversely affecting the marine ecological environment. Furthermore, forward osmosis (FO), which utilizes natural osmosis to eliminate the need for pressurization, has become an emerging technology, but bottlenecks in high-efficiency membrane design have hindered its application.

In this context, membrane distillation (MD) has emerged as a promising seawater desalination technology due to its advantages of low energy consumption, high salt rejection and ability to handle high-concentration brine [4]. As shown in **Figure 1**, MD is a thermal-driven process that utilizes a hydrophobic microporous membrane as a barrier between hot-side feed solution and the cold-side permeate solution. The vapor pressure difference between the two sides of the membrane is the mass transfer driving force, which separates non-volatile components [5]. Membrane, as the core of MD, plays an important role in its operation. Traditional hydrophobic membranes are prone to fouling and wetting, which impedes the industrial application of MD. Fortunately, super liquid-repellent membranes inspired by natural surfaces, such as lotus leaves, fish scales, and cicada wings, can alleviate membrane fouling and enhance membrane wetting performance. Superhydrophobic membranes, omniphobic membranes, and underwater superoleophobic membranes fall under this category.

The fabrication of hydrophobic membranes often employs non-solvent induced phase separation (NIPS), thermally induced phase separation (TIPS), sintering, and track-etching [6]. However, electrospinning technology, which can produce nanofibrous membranes with unique structural features, is superior to these methods. Electrospinning technology can construct super liquid-repellent membranes with concave geometries and facilitate the large-scale production of nanofibrous membranes. Additionally, incorporating functional polymers or nanoparticles in spinning precursors or combining electrospinning with membrane surface modification can create super liquid-repellent membranes with different wettability.

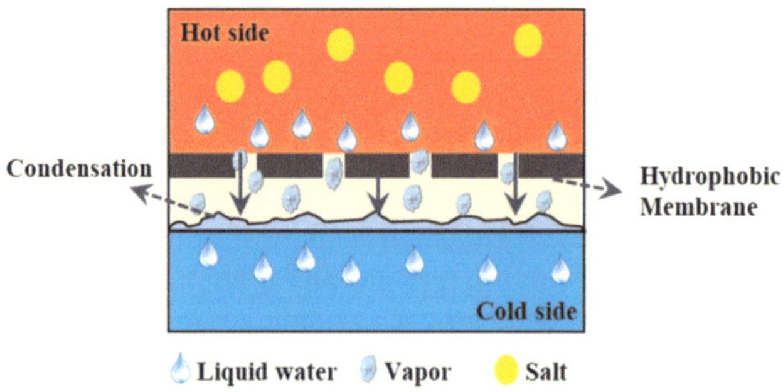

Figure 1.
Schematic diagram of membrane distillation process.

This chapter provides a review of recent progress in the fabrication of super liquid-repellent electrospun nanofibrous membranes (ENMs) using electrospinning technology. It discusses the electrospinning mechanism and influencing parameters, followed by the theory and preparation of super liquid-repellent membranes. The section that follows illustrates the applications and challenges of these membranes in MD. Finally, the paper concludes with a brief summary and outlook.

2. Theory of membrane distillation

2.1 MD configurations

As mentioned above, MD is a thermal-driven membrane process in which the water vapors transport through a microporous hydrophobic membrane and condense on the cold permeate side [4]. Theoretically, the membrane's inherent hydrophobicity ensures its non-wettability by the feed solution. On the thermal boundary, the hot-side feed directly engages the hydrophobic membrane surface, effectuating evaporation at the membrane interface. The motion of steam across the membrane is impelled by pressure differentials, leading to its condensation on the permeate side. The defining attributes of MD technology encompass the subsequent elements [7]: (1) the membrane is hydrophobic and microporous, (2) at least one side of the membrane surface is in contact with the feed liquid and will not be wetted, (3) conversion of volatile feed liquid constituents into steam, transiting through membrane pores, and (4) exploitation of the vapor pressure gradient across the membrane to propel the passage of volatile constituents through its pores.

To counteract the thermal losses stemming from membrane conduction during MD operations, various MD configurations have evolved. Conventionally, there are four designs for MD configurations found in the literature [8]: Direct contact membrane distillation (DCMD), air gap membrane distillation (AGMD), vacuum membrane distillation (VMD) and sweeping gas membrane distillation (SGMD). A compendium of the principal attributes, merits, and demerits of these four MD configurations is tabulated in **Table 1**.

2.2 Methods of fabrication of MD membranes

Hydrophobicity and porosity constitute pivotal attributes of MD membranes. Various techniques, such as phase inversion, stretching, track-etching, and electrospinning, are conventionally employed for the fabrication of diverse MD membranes [9]. In addition, different forms of membranes in MD such as single-layer membranes and composite membranes are used in different MD scenarios to enhance MD performance.

The phase inversion process hinges upon a transition between two phases, induced by altering the solubility of the polymer constituents [10]. Through modulation of the composition or conditions, the initially uniform polymer solution undergoes phase separation, leading the casting solution's transition from a liquid to a solid state. The ultimate membrane is acquired by solidifying the polymer-rich phase. Versatile membrane structures are typically realized via non-solvent induced phase separation (NIPS), temperature induced phase separation (TIPS), and vapor induced phase separation (VIPS). It is worth noting, however, that the viable membrane materials are constrained due to the necessity for soluble polymers in the phase inversion process.

MD configuration	Main characteristics	Advantages	Disadvantages
DCMD	One side of the hydrophobic membrane is in contact with the hot side feed solution and the other side is in contact with the cold side permeate solution.	The simplest configuration. Easy operation. High water flux.	Lower thermal efficiency. Highly affected by temperature polarization
AGMD	The hydrophobic membrane is only in contact with the hot side feed solution. An extra air gap compartment is introduced between the other side of the membrane and the condensation plate.	Less heat loss.	Greater mass transfer resistance
VMD	An additional vacuum pump is installed on the condensing side, and the permeated vapor is condensed in an external condensing device.	High flux. Heat loss can be ignored.	Membrane wetting is prone to occur. Complex configuration.
SGMD	The inert gas introduced on the permeation side will carry the vapor passing through the membrane pores to the external condensing device for condensation.	Greater mass transfer drive. Lower mass transfer resistance compared to AGMD.	High energy consumption. Complex configuration.

Table 1.
The main attributes, merits, and demerits of the four MD configurations.

Stretching represents a solvent-free approach to membrane fabrication, entailing the heating of a polymer beyond its melting point and extruding it into the desired form under elevated stress, thereby engendering membrane pores through elongation [9]. This methodology finds suitability in the case of highly crystalline polymeric materials, such as Polytetrafluoroethylene (PTFE) and polyethylene (PE). Notably straightforward in terms of equipment and procedural intricacy, this technique exhibits environmental merits. Nevertheless, the resultant thin membrane is somewhat circumscribed by its substantial pore size and broad pore size distribution.

Track-etching is a widely employed means to fabricate membranes with small pore sizes by irradiating non-porous polymer membranes with high energy heavy ions to create linear damaged tracks [11]. Accurate control of the pore size distribution can be achieved using this method.

These methodologies serve the purpose of scalable MD membrane production, however, substantial technical intricacies still warrant comprehensive exploration. These challenges encompass issues such as restricted permeation flux and the propensity for facile pore wetting. These concerns emanate from the inherent deficiencies in membrane surface hydrophobicity and surface roughness, alongside the constrained porosity, diminutive pore dimensions, and the presence of intricate tortuous closed-pore configurations within the membrane matrix. Conversely, nanofiber membranes produced via the intricate art of electrospinning present a contrasting picture. These membranes feature an intricate network of interconnected pores, distinguished by elevated porosity levels, and underscored by the added advantage of precise control over fiber morphology and membrane thickness. Notably, the process for preparing these nanofiber membranes is streamlined and straightforward, further enhancing their appeal. As a result, within the realm of membrane distillation, researchers have increasingly directed their focus toward these ENMs, recognizing their potential, as evidenced by the mounting scholarly interest in recent years.

3. Theory of electrospinning process

Electrospinning is a highly versatile and cost-effective technique for fabricating nanofibrous membranes with exceptional specific surface area, pore structure, and surface morphology. This technology was first discovered in 1897 and made significant progress in the early 1990s after undergoing a series of detailed studies [12]. Nanofibrous membranes produced by electrospinning typically exhibit high porosity and interconnected pores, which reduce mass transfer resistance within the structure, making them highly suitable for membrane separation applications. In this section, we will discuss the fundamental principles of electrospinning technology and the influence of various parameters on the nanofibrous membrane during the spinning process.

3.1 Understanding electrospinning

In recent years, considerable attention has been given to the production, characterization, and application of electrospun nanofibers, as well as the enhancement of spinning devices for more precise and efficient fabrication of complex nanofibers tailored to specific applications. To this end, various sophisticated electrospinning devices have been developed, such as vibration electrospinning, magneto-electrospinning, and melt electrospinning [13]. Nevertheless, regardless of the design, every electrospinning apparatus is composed of four fundamental

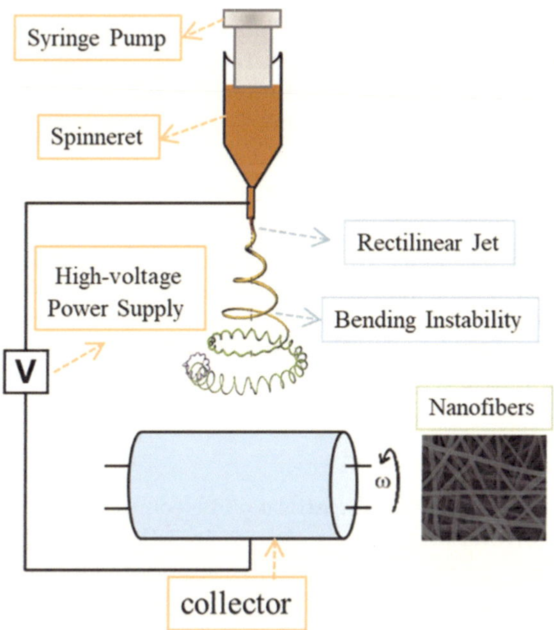

Figure 2.
Schematic diagram of electrospinning process.

components: a high-voltage power supply, a syringe pump, a spinneret, and a collector (**Figure 2**) [14]. Generally, a high-voltage power supply refers to a DC power supply capable of generating thousands to tens of thousands of volts, which will form a high-voltage electric field to charge the fluid, causing disturbance and polarization of the fluid, and finally forming a jet. The spinneret is connected to a container containing the polymer solution, and the fluid flows through the spinneret at a constant and controllable speed driven by the syringe pump. The collector, typically constructed from metal, is either a rotating drum or aluminum foil connected to the negative terminal of the high-voltage power supply.

Electrospinning is a distinctive technique of producing fibers utilizing electrostatic forces. At room temperature and atmospheric conditions, a polymer solution or molten liquid is placed under a high-voltage electric field to form a charged jet. As the jet traverses the air, solvent evaporation and fiber solidification occur, leading to the formation of polymeric nanofibers on the collecting plate [15]. The electrospinning process typically progresses through three primary phases: (1) spindle formation and jet initiation; (2) rectilinear jet development and bending instability, and (3) nanofiber solidification and collection.

Initially, a high-voltage power supply is employed to apply a high constant electric field between the polymer droplets at the tip of the spinneret and the grounded collector, charging the pendulous droplet at the spinneret. Subsequently, through the action of Coulombic repulsion force and electrostatic force, the semicircular droplet progressively changes into a cone shape, known as the Taylor cone. Above a critical electric field, the electrostatic force surpasses the surface tension of the polymer solution, resulting in the ejection of a fine jet from the tip of the Taylor cone.

During flight, the jet first moves in a straight line at a distance from the spinneret that has a critical value proportional to the applied electric field, electrical conductivity and flow velocity of the polymer solution [14]. After linear motion, electrically induced bending instabilities occurs, forming a conical unstable region, which is crucial for refining the fiber diameter from micrometers to nanometers. Ultimately, rapid solvent evaporation due to the high surface area to volume ratio leads to the solidification of the fibers and the production of the desired nanofiber membrane on the collection plate.

3.2 Different parameters affecting resultant nanofibrous membranes

Different parameters in the electrospinning process including the inherent properties of the polymeric dopes, operational conditions and surrounding environmental conditions can significantly affect the resultant nanofibrous membranes. By controlling these parameters, nanofibers with ideal morphology and diameter can be obtained.

3.2.1 Effect of polymeric dopes

It is widely acknowledged that the efficacy of electrospinning is significantly influenced by the polymer materials employed [16, 17]. In the MD process, the materials that can be used to manufacture nanofibrous membranes need to be hydrophobic and soluble. Different types of polymer materials have different molecular weights, which play a pivotal role in determining the electrical and rheological characteristics of the spinning solution. Generally, smaller molecular weights lead to lower solution viscosities, resulting in smaller fiber diameters that are beneficial for mitigating wetting but not permeate fluxes. Furthermore, the diameter of the nanofibers depends greatly on the concentration of the polymer solution. When the concentration of the solution is too low, nanofibers with a beaded structure are formed, and when the concentration is increased, the spherical nanofibrous microspheres tend to become spindle-like shapes [18].

During the spinning process, a high-voltage electric field is required for the polymer solution to be ejected from the end of the spinneret [19]. Hence, it is imperative that the solution is conductive to a certain extent to ensure that it can be ejected appropriately. If the solution's conductivity is too strong, the ejected fiber will undergo considerable bending deformation, which is not conducive to its spreading on the collector, resulting in microbeads and a dense network [20]. Conversely, too low conductivity results in insufficient stretching of the jet, leading to non-uniform fibers. Therefore, at the initial stage of spinning, it should be ensured that the conductivity of the polymer solution is within an appropriate range. Usually, ions are introduced into the spinning solution to adjust the conductivity and increase the spinnability due to their high mobility and charge density [20].

It was revealed that the morphology and properties of electrospun nanofibers are affected by the surface tension of polymer solutions [21, 22]. A suitable surface tension contributes to the stability of the jet. By reducing the surface tension of the solution, the electrospinning process can be completed at lower electric field strength. Adding a suitable dose of surfactant in the electrospinning solution helps to generate uniform nanofibers. However, an excessive amount of surfactant can lead to self-assembly, causing colloidal aggregation and ultimately resulting in defects on the fiber surface.

The choice of solvent is crucial because its evaporation governs the solidification of nanofibers during electrospinning. The solvent type affects the solution's viscosity, with higher viscosities making it difficult for polymer solutions to eject under appropriate electric field strength and lower viscosity producing discontinuous fibers. Currently, N, N-dimethylformamide (DMF) and N-methylpyrrolidone (NMP) are relatively common polar solvents that can dissolve polymers, while acetone is a non-solvent that can dilute polymer solutions. Usually, both solvent and non-solvent are utilized to prepare the spinning solution.

3.2.2 Effect of operational conditions and surrounding environmental conditions

In addition to the intrinsic properties of polymer solutions, extrinsic factors such as operational and environmental conditions can also exert a substantial influence on the nanostructure of ENMs. Among the operational parameters, spinning time directly dictates the thickness of the nanofibrous membrane and hence modulates the membrane permeability. The working distance, which denotes the distance between the spinneret and the collector, affects the strength of the electric field and the evaporation kinetics of the solvent. An excessively short working distance may impede solvent evaporation and lead to undesired adhesion among fibers. Another critical parameter is the applied electric field. Theoretically, a higher voltage facilitates fiber stretching and formation of smaller diameter fibers. However, as the voltage increases, a "frying" phenomenon may occur, reducing the collection efficiency of fibers [6]. In actual operation, the voltage of the spinning system should be adjusted according to the shape of the "Taylor cone."

Apart from operational factors, the ambient environmental conditions, such as temperature and relative humidity, can also significantly affect the nanostructure of fibers [23]. Temperature has a dual effect on the viscosity and surface tension of the spinning solution and influences the evaporation rate of the solvent. A higher temperature leads to a lower viscosity and surface tension, resulting in smaller diameter nanofibers. Furthermore, it can promote solvent evaporation and facilitate fiber solidification. Conversely, relative humidity is an essential factor that can impact the evaporation kinetics of the solvent, thereby affecting the resulting nanofibrous membranes. A low relative humidity can accelerate the evaporation rate and cause premature coagulation of the spinning solution, which may clog the spinneret and interfere with the spinning process. Therefore, it is crucial to control the humidity to achieve the desired nanofiber membrane morphology.

4. Preparation of super liquid-repellent surfaces by electrospinning

In the context of mitigating membrane wetting and fouling during the MD process, the development of super-liquid repellent membranes with exceptional wetting properties has emerged as a current research focus. Compared with other fabrication methods, electrospinning is a superior technique for producing nanofibers with good functionality, high porosity, and specific surface area. ENMs have gained extensive utilization in MD due to their low energy consumption and high efficiency. By means of physical and chemical modifications, the interfacial properties of ENMs can be optimized to prepare super-liquid repellent membranes that are suitable for treating complex wastewater in MD. This section will elaborate on the preparation of super-liquid repellent membranes based on electrospinning, in accordance with the classical wetting model.

4.1 Superhydrophobic surfaces

4.1.1 Theory of superhydrophobic membranes

Highly hydrophobic membranes can effectively prevent wetting by aqueous solutions during the MD process. Superhydrophobic surfaces have good resistance to wetting, which is defined as water contact angle (WCA) greater than 150° [24]. Inspired by the "lotus leaf effect," it was found that the self-cleaning properties of the surface were attributed to the combination of micron-structured papillae and epicuticular wax layers [25]. Subsequent studies suggested that there are nanostructures on the surface of the lotus leaf, and the combination of micro/nanostructures together promotes its superhydrophobicity and self-cleaning properties. This discovery provides a theoretical basis for the construction of superhydrophobic surfaces.

The interfacial tension between the contacting phases determines the degree of wetting of the membrane surface, which is usually characterized by the contact angle (CA), the larger the CA, the less droplets adhere to the membrane. Young's equation is the simplest equation to describe the wettability of a surface, and is usually used to describe the state of a water droplet on an ideal flat (a smooth and chemically uniform surface) [26, 27]:

$$\cos\theta_{eq} = \frac{\gamma_{SG} - \gamma_{SL}}{\gamma_{LG}} \quad (1)$$

where θ_{eq} is the CA of an ideal solid surface. γ_{SG}, γ_{SL} and γ_{LG} represent the interfacial tension between solid–gas, solid–liquid and liquid–gas, respectively.

According to this equation, it can be known that the CA of the surface can be changed by introducing different chemical groups [28].

However, perfectly smooth solid surfaces rarely exist in practical situations. Therefore, the Wenzel is used to describe the wettability of rougher surfaces [29]:

$$\cos\theta_W = r\cos\theta_{eq} \tag{2}$$

where r is the roughness factor, defined as the ratio of the wetted surface to the projected surface. θ_W represents the apparent WCA of the rougher surface. According to this theory, the wettability of surfaces with greater surface roughness will be enhanced. The Wenzel model is further extended by the Cassie-Baxter theory, which assumes that a rough porous surface will not be filled with liquid due to the presence of air pockets [29, 30]:

$$\cos\theta_{CB} = f_{sl}\left(1 + \cos\theta_{eq}\right) - 1 \tag{3}$$

where θ_{CB} is the apparent contact angle under this theoretical model and f_{sl} is the contact area fraction of solid–liquid. According to this theory, a superhydrophobic surface can be obtained by reducing the solid–liquid interface. The left side of **Figure 3(A–C)** shows the wetting behavior of water droplets in three different states.

4.1.2 Engineering in-air superhydrophobic surfaces by electrospinning

Based on the aforementioned theory, superhydrophobic surfaces are constructed by designing surfaces with a certain roughness or using low surface energy materials. So far, studies have successfully combined bionics-based electrospinning technology with various modification procedures to prepare superhydrophobic surfaces for MD, including: one-step electrospinning, surface modification and mixed matrix methods.

One-step electrospinning involves mixing low surface energy materials into the spinning solution, which is a straightforward and effective approach to generate superhydrophobic nanofiber membranes. For instance, perfluorooctyltriethoxysilane

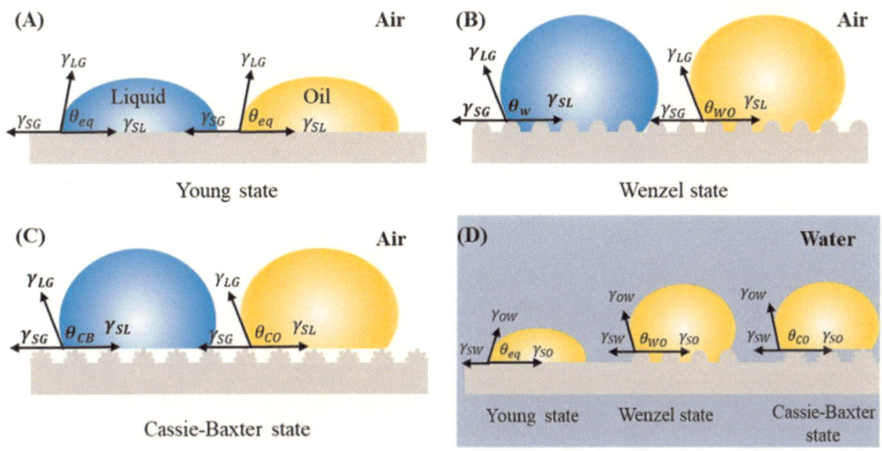

Figure 3.
Typical wetting behavior of a droplet on the solid substrates. (A) in-air Young's model. (B) in-air Wenzel's model. (C) in-air Cassie's model. (D) underwater Young's model, Wenzel's model and Cassie's model.

(PFOTES) has been added to polymer solutions to synthesize fluorinated silane functionalized polyvinylidene fluoride (PVDF), and the nanofiber membranes obtained by electrospinning using this solution exhibited a higher WCA than those prepared by solvent evaporation (**Figure 4**) [31]. Furthermore, the ENMs without PFOTES (**Figure 4B**) contained a large number of beaded structures, while the formation of bead-free nanofibers imparted a stronger hydrophobicity to the PVDF/PFOTES nanofiber membranes (**Figure 4C**). PTFE is also selected to be added to the PVDF polymer solution for electrospinning. Within a certain range, the WCA of the membrane increased with increasing PTFE concentration, gradually exhibiting superhydrophobicity [32].

Despite the simplicity of one-step electrospinning, it has certain limitations since the available polymer materials for direct electrospinning are relatively limited. Therefore, it is necessary to explore the sources of hydrophobic materials further. In recent years, many attempts have been made to directly deposit low surface energy materials on the surface of ENMs to prepare superhydrophobic membranes suitable for MD processes. Chemical vapor deposition (CVD) has been utilized to produce superhydrophobic membranes. MD membranes with a WCA of 151° can be created by depositing poly (1H,1H,2H,2H-perfluorodecyl acrylate) (PPFDA) on an ENMs substrate. In this study, a custom-built reactor was used for deposition of PPFDA [33]. In addition, low surface energy fluoroalkylsilane (FAS) was used to reduce the surface tension of the membrane. The prepared electrospun nanofibrous membrane was placed in FAS solution (2 wt. %) for hydrophobic modification process. The resulting nanofiber membrane F-PVA can reach a water contact angle of 158° [34]. In a preceding investigative endeavor from our body of work, we detailed the development of a superhydrophobic membrane designated as PVDF-S, distinguished by its heightened resistance to wetting. This innovative membrane was demonstrated to exhibit a sustained and stable performance within the context of MD, particularly when subjected to feed solutions containing DTAB (N,N,N-trimethyl-1-dodecanaminium bromide). The foundational construction of the PVDF substrate membrane itself was realized via the intricate process of electrospinning, thereafter subjected to fluorination through the application of a low surface energy FAS. The ensuing fluorination procedure, integral to our approach, effectively brought about a substantial reduction in surface energy, consequently conferring a notable enhancement in hydrophobic characteristics upon the membrane [4].

Surface modification, which reduces the surface energy of the membrane, can moderately increase hydrophobicity. However, the roughness of the nanofibers produced by electrospinning is inadequate, which limits further enhancement of the final membrane hydrophobicity. According to the theory of Wenzel and Cassie-Baxter, in addition to reducing the surface energy, another way to achieve superhydrophobic

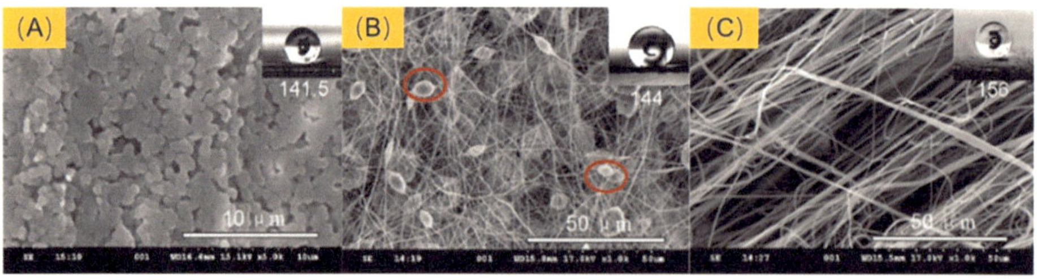

Figure 4.
SEM images of membranes prepared from (A) PVDF/PFOTES using solvent evaporation, (B) PVDF, (C) PVDF/PFOTES using electrospinning. Reproduced from [31] with permission. Copyright (2009) Elsevier Inc.

Membrane name	Category	WCA (°)	OCA (°) In-air/underwater		SA (°)
PVDF/PFOTES	Superhydrophobic	156	/	/	/
PVDF-PTFE	Superhydrophobic	152.2 ± 2.0	/	/	/
PPFDA-iCVD	Superhydrophobic	151 ± 2	/	/	/
F-PVA	Superhydrophobic	158	/	/	4
PVDF-S	Superhydrophobic	>150	>120	<30	36
PVDF/SiO_2	Superhydrophobic	157 ± 1	/	/	3.7 ± 1.2
P/CF-60	Omniphobic	160.9 ± 0.9	~149°	/	~51
SiNPs-PVDF-HFP/BTEAC	Omniphobic	~150	~139	/	/
F-POSS/PVDF-HFP	Omniphobic	154.5 ± 2.6	148.8 ± 3.7	/	/
FAS@SiNPs-SFM	Omniphobic	154.2 ± 3.6	154.2 ± 3.6	/	/
PVDF-F	Omniphobic	164.5 ± 1.7	>150	/	1.6 ± 0.1
Janus(o)	Underwater superoleophobic	<30	<60	~149°	/
PH-7.5RC	Underwater superoleophobic	0	0	140.5 ± 2.62	/
PTFE/PAN-OH	Underwater superoleophobic	<40°	/	161.7 ± 1.8	/

Table 2.
The experimental values of water contact angle (WCA), oil contact angle (OCA), and sliding angle (SA) for superhydrophobic membranes, omniphobic membranes, and Janus membranes.

properties is the construction of micro/nanoscale rough structures. The mixed matrix method refers to combining multi-level roughness and low surface energy surface through post-treatment or doping of bulk solution to prepare superhydrophobic membranes. Typically, various nanoparticles such as silicon dioxide (SiO_2), titanium dioxide (TiO_2), and silver nanoparticles (AgNPs) are usually applied to increase the roughness of the membrane surface [5, 35, 36]. For example, hydrophobically modified SiO_2 was incorporated into the spinning solution for one-step electrospinning to construct MD membranes with outstanding properties, and its WCA reached 157°. Compared with the unmodified PVDF membrane, the water osmotic pressure of the superhydrophobic membrane doubled, and the water flux was close to 25.7 kg m^{-2} h^{-1} during 100 h of DCMD [37]. **Table 2** summarizes the experimental values of water contact angle (WCA), oil contact angle (OCA), and sliding angle (SA) for superhydrophobic membranes, omniphobic membranes, and Janus membranes mentioned in this chapter.

4.2 Omniphobic surfaces

4.2.1 Theory of omniphobic membranes

Despite the ability of superhydrophobic membranes to alleviate wetting and fouling, their practical operation can be hampered by complex feed solutions, which can easily induce membrane wetting and halt the MD process. Young's equation predicts that reducing surface tension enhances wettability, whereas Wenzel's theory proposes that increasing the surface roughness of the membrane improves its anti-wetting ability. Furthermore, Cassie-Baxter theory suggests that reducing the contact area between solid and liquid is a method to further increase liquid repellency. Therefore, omniphobic membranes with re-entrant structures and low surface energy surfaces is suitable for treating complex feed liquids. These membranes are both hydrophobic and oleophobic, allowing them to retain both polar and non-polar liquids. A good omniphobic membrane should possess a relatively high water and oil contact angle, high flux, high LEP value, and excellent long-term stability.

As illustrated in the right side of **Figure 3(A–C)**, Young's, Wenzel's and Cassie-Baxter's models are also applicable in describing the wettability of oil droplets on flat or rough solid surfaces. In this context, the liquid in the model refers to oil droplets. When in the Wenzel state, the oil droplets with low surface tension can easily wet the surface of the ridged structure. The wetting tendency increases with increasing distance between two ridges. In contrast, when in the Cassie-Baxter state, due to the presence of the re-entrant structure and the low surface energy surface, the air below the recessed structure generates a negative Laplace pressure difference, which plays a key role in preventing further penetration of the oil droplets [28].

4.2.2 Engineering in-air omniphobic surfaces by electrospinning

To construct an omniphobic membrane surface, two key factors must be considered: (1) building a re-entrant structure to enhance the surface roughness, and (2) creating a surface with low surface energy. The preparation of omniphobic membranes via electrospinning involves three main techniques. The first method involves electrospinning a polymer-only solution, followed by modifying the resulting fibers with low surface energy materials or nanoparticles. The second approach involves adding fluoride to the electrospinning solution, resulting in the formation of low surface energy nanofibers. Finally, the third method involves incorporating nanoparticles into the spinning solution to create roughened nanofibers, which are then treated with fluoride.

The first method is commonly used to achieve omniphobic membrane surfaces. Plasma treatment is an effective technique to enhance omniphobicity. For instance, Woo et al. formed new CF_2-CF_2 and CF_3 bonds on the surface of ENMs, reducing their surface energy and ensuring stable MD performance of the prepared omniphobic membranes P/CF-60 for real RO brine [38]. They modified ENMs using a plasma system equipped with parallel plate electrodes coupled with a radio frequency (RF) glow discharge system. ENMs were placed on the plates and CF_4 gas was introduced into the chamber at a certain flow rate. Plasma treatment can also generate in-plane nanopores in the nanosheets and provide active sites for constructing omniphobic membranes with stable performance [39]. Dip-coating is another method used to design ENMs with excellent wetting properties. Lee et al. added cationic surfactants to the PVDF-HFP spinning solution, making the substrate positively charged. They then grafted negatively charged SiNPs onto the surface by dip-coating and fluorinated it, obtaining omniphobic membranes with a low surface energy surface and a concave structure. The prepared membranes SiNPs-PVDF-HFP/BTEAC showed stable performance in the 8 h DCMD test [40].

The second way is to obtain omniphobic membranes by directly adding fluoride to the spinning solution. Omniphobic membranes with uniform fibrous structure can be obtained by adding fluorinated-decyl polyhedraloligomeric silsesquioxane (F-POSS) to the PVDF-HFP suspension solution, which exhibits good wetting resistance to liquids with different surface tensions and shows stable MD performance (**Figure 5**) [41].

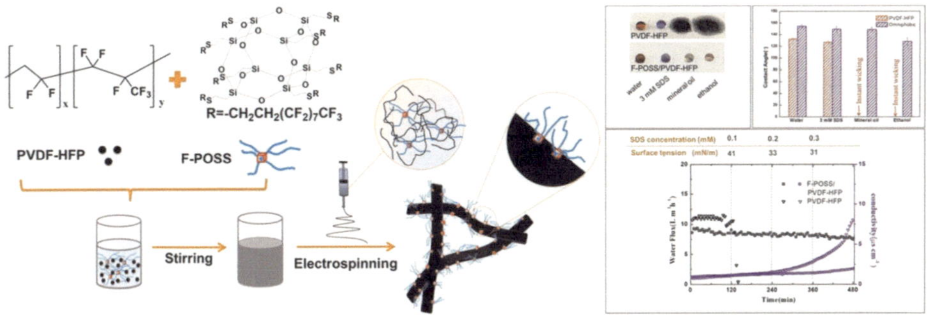

Figure 5.
Schematic diagram of preparation method, wetting performance, and MD performance of the omniphobic membrane. Reproduced from [41] with permission. Copyright (2018) Elsevier Inc.

Similarly, Lu et al. also used F-POSS as a fluorinating agent to prepare an omniphobic membrane with a contact angle of 120.3° to ethanol. Spinning solutions were prepared by dispersing F-POSS particles (3 wt.%) in a mixed solution containing PVDF-HFP, DMF, and acetone, followed by electrospinning [42].

The last method involves incorporating nanoparticles into the spinning solution. Huang et al. used SiNPs to prepare a sheath solution for coaxial electrospinning to produce fibers with nanoscale roughness, and a PVA/silica solution as the core solution. After fluorination by FAS, a hydrophobic and oleophobic membrane was obtained, which exhibited resistance to surfactants in MD tests [43]. Generally, SiNPs, AgNPs, and TiO_2 are often used to increase the surface roughness of the membrane. In our previous study, a robust omniphobic membrane #PVDF-F is developed by fluorinating silver nanoparticles (AgNPs)-coated nanofibrous membrane, which has a concave structure with low surface energy [5].

4.3 Underwater superoleophobic surfaces

4.3.1 Theory of underwater superoleophobic membranes

Janus membranes with asymmetric wettability usually have underwater superoleophobic surfaces and are widely used for treating oily wastewater in MD. The term "Janus" is derived from ancient Greek and Roman mythology and refers to a god with two faces. As shown in **Figure 6**, Janus membranes for MD are usually composed of a hydrophilic surface layer and a substrate with different hydrophobicity. Thanks to their asymmetric properties, properly designed Janus membranes exhibit resistance to wetting and fouling when faced with different types of liquids.

The underwater oleophobicity of the surface layer is an important factor that influences the performance of Janus membranes. Similarly, the three classical wetting models mentioned above are applicable to the wetting behavior of underwater oil droplets on solid surfaces. The underwater oil contact angle (UOCA) for the ideal case can be calculated using Eq. (4) [44]:

$$\cos\theta_{eq^*} = \frac{\gamma_{SO} - \gamma_{SW}}{\gamma_{OW}} = \frac{\gamma_{OA}\cos\theta_{OA} - \gamma_{WA}\cos\theta_{WA}}{\gamma_{OW}} \tag{4}$$

where, γ_{SO}, γ_{SW} represent the interfacial tension of solid-oil and solid-water, γ_{OW} is the surface tension of oil–water, and θ_{eq^*} refers to the CA of oil droplets under water.

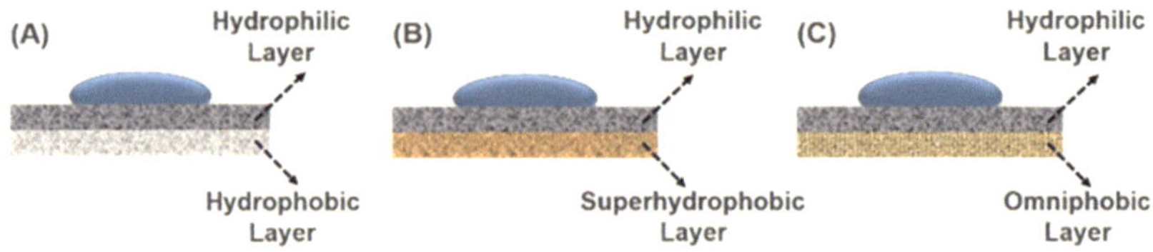

Figure 6.
Schematic diagram of three common Janus membranes.

Similarly, when medium to large-scale roughness exists on solid surfaces, oil droplets exist in the Wenzel state and the underwater equation is [44]:

$$\cos\theta_{UWO} = \mathrm{r}\cos\theta_{eq^*} \tag{5}$$

where θ_{UWO} is the UOCA on a solid surface in the Wenzel state, and the roughness of the solid surface determines its magnitude. In contrast, in the underwater Cassie-Baxter state, the grooves of the hierarchical micro/nanostructure on the rough solid surface are filled with water, which endows it with underwater oleophobicity and effectively reduces the adhesion of oil droplets to the membrane surface during MD. The equation in this state is [44]:

$$\cos\theta_{UCO} = \mathrm{f}_{SO}(1 + \cos\theta_{OW}) - 1 \tag{6}$$

where f_{SO} and θ_{UCO} are the contact area fraction and UOCA of the multi-level roughness solid surface, respectively. **Figure 3D** shows the wetting behavior of oil droplets on different membrane surfaces under water.

4.3.2 Engineering underwater superoleophobic membranes by electrospinning

The underwater superoleophobic top layer of the Janus membrane can form a hydration layer at the solid–liquid interface, effectively inhibiting the adhesion of oil droplets on the membrane surface. Therefore, its preparation should consider the surface morphology and hydrophilic chemical functional groups of the top layer. Additionally, due to the requirement of hydrophobicity in the MD process, substrate modification should be fully considered. There are two main ways to prepare Janus membranes using electrospinning technology: the first method is to use the combination of electrospinning technology and post-modification to prepare a double-layer composite membrane; the second method is to uses electrospraying technology to introduce hydrophilic materials on the hydrophobic substrates.

At present, some researches have realized the successful preparation of Janus membrane through post-treatment modification based on electrospinning. For instance, electrospinning was used to prepare positively charged CTAB/PVDF-HFP substrates, followed by adsorption of negatively charged SiNPs on it by dip-coating method, and finally fluorination (**Figure 7A**). The prepared membranes had a secondary re-entrant structure (**Figure 7B**) and exhibit omniphobicity. Finally, SiNPs, chitosan (CTS), and perfluorooctanoate (PFO), SiNPs-CTS/PFO, was sprayed on the substrate to obtain Janus membranes Janus (O). Wettability tests revealed that the membrane Janus (O) has asymmetric wettability and its top layer is underwater oleophobic (**Figure 7C**) [45]. In addition, the simple stacking of two membranes with different wettability based on electrospinning can also obtain a dual layer membrane

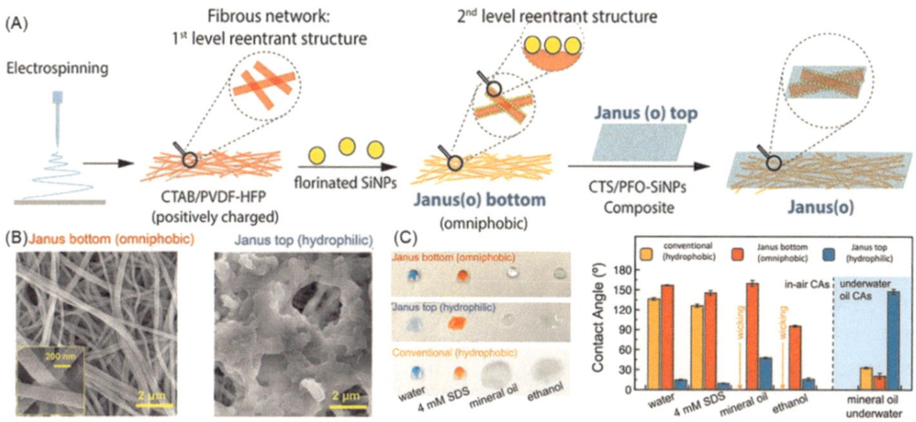

Figure 7.
(A)Schematic diagram of fabrication procedure of the Janus membrane, (B) SEM image of the substrate and surface of Janus membrane, (C) photographic images and CA of different liquid droplets on the bottom surface and top surface of the Janus membrane. Reprinted (adapted) with permission from [45]. Copyright (2017) American Chemical Society.

with underwater oleophobicity. Makanjuola et al. directly coated different concentrations of cellulose on PVDF-HFP membranes to obtain Janus membranes that can treat oily wastewater [46]. Vacuum filtration, UV-mediated modification strategy, multi-step coating and surface modification can usually achieve the preparation of the Janus membrane hydrophilic layer [47]. Our previous study prepared Janus membranes with hydrophobic bottom and hydrophilic top by spray-coating multi-walled carbon nanotubes (MWCNTs) on electrospun PVDF substrate, which showed improved anti-fouling properties [4].

Most studies use commercial membranes as substrates and use electrostatic spraying of hydrophilic materials to prepare Janus membranes. Hydrophilic materials such as CTS, polyacrylonitrile (PAN), polydopamine (PDA), polyethyleneimine (PEI) and carbon nanotubes (CNT) are usually used for the preparation of the underwater oleophobic layer of Janus membranes [48–51]. For instance, Tang et al. prepared Janus membranes with asymmetric wetting characteristics by electrospraying PAN on the surface of hydrophobic PTFE membrane and hydrolyzing it with ethylenediamine (EDA) and sodium hydroxide (NaOH). The Janus membranes PTFE/PAN-OH exhibited a UOCA > 150° and displayed stable permeate flux and salt rejection in MD-treated oil-containing feed solution [48]. Overall, so far, the research on the preparation of Janus membranes by electrospinning/spraying has become increasingly mature, which is of great significance to the industrial application of MD technology.

5. Application of super liquid-repellent membrane in MD

In comparison to conventional hydrophobic membranes, superhydrophobic and omniphobic membranes offer superior performance in the MD process for the treatment of complex components and contaminants owing to their robust wetting stability. Superhydrophobic membranes are widely utilized for seawater desalination and brine treatment due to their excellent permeability. When dealing with high-concentration brine (25 wt%), the superhydrophobic membrane maintained a stable flux and salt rejection in the 180 h DCMD test [52]. Notably, superhydrophobic membranes also demonstrated remarkable long-term stability in treating industrial wastewater [53–55].

However, due to limited hydrophobicity, superhydrophobic membranes are not suitable for wastewater containing a variety of low surface tension substances. Omniphobic membranes, on the other hand, possess stronger hydrophobicity, which imparts exceptional repellency to various low surface tension liquids and enables them to resist wetting more effectively than superhydrophobic membranes. Therefore, many studies have highlighted the effectiveness of omniphobic membranes in treating low surface tension wastewater [5, 56, 57]. The modified omniphobic membrane exhibited excellent wetting resistance to 0.6 mM sodium dodecyl sulfate (SDS) [58]. In addition, the interactions between surfactants with different properties, such as anionic surfactant SDS, dodecyltrimethylammonium bromide (DTAB), and nonionic surfactant polyoxyethylenesorbitan monolaurate (Tween-20), and omniphobic membranes were elucidated [4]. Interestingly, MD membranes with omniphobicity can also treat wastewater containing emulsified oil. Specifically, the omniphobic membranes showed stable performance in treating Oil-emulsion containing 0.005% v/v [59].

Janus membranes, which feature an underwater oleophobic layer, have emerged as a promising candidate for the treatment of wastewater containing low surface tension chemicals and oily wastewater. Our group investigated the MD performance of Janus membranes with different substrates (hydrophobic, superhydrophobic, and omniphobic), and found that Janus membranes with omniphobic substrates had the best stability for the treatment of surfactants and oil-in-water emulsions [4]. The hydrophilic surface layer of the Janus membrane can form a hydration layer at the solid–liquid interface to prevent the adhesion of oil droplets to the membrane surface. In addition, the size exclusion of the hydrophilic layer proved to be effective for surfactants [60].

In conclusion, super liquid-repellent membranes have broad application prospects in the fields of desalination, wastewater treatment, and oil–water separation due to their excellent wettability, which is beneficial to promote the industrial application of MD technology.

6. Challenges facing super liquid-repellent membrane based on electrospinning in MD

Compared with traditional hydrophobic membranes, although super liquid-repellent membranes prepared by electrospinning have significant advantages, they also face many challenges that hinder their commercial application. Firstly, the fabrication process of the super liquid-repellent membrane remains intricate, and the membrane modification materials employed, such as fluorinated solvents, are both expensive and potentially toxic, which could pose hazards to drinking water production. Secondly, despite the laboratory research indicating the efficacy of super liquid-repellent membranes, there remains a paucity of practical application in large-scale production. Finally, the balance between modified membrane permeability and porosity must be attained, and the mechanical robustness of the Janus membrane requires continuous improvement. As such, it is imperative to conduct research on more eco-friendly, stable, and producible super liquid-repellent membranes and to investigate the fouling and wetting mechanisms of real wastewater in MD to expand their application range.

7. Conclusions and outlooks

ENMs present promising prospects for future MD applications as the next generation of MD membranes, owing to their exceptional porosity, hydrophobicity, and feasibility in constructing multilayered membranes. The utilization of electrospinning-based super liquid-repellent membranes in MD for anti-wetting and anti-fouling purposes has garnered significant interest. Over the past few years, the amalgamation of innovative nanomaterials and various modification strategies with electrospinning has yielded super liquid-repellent membranes with specific morphology and higher permeation flux, significantly enhancing the competitiveness of ENMs in treating low surface tension wastewater. Recent findings suggest that improved nanofibrous membranes are promising options for applications in seawater desalination, brine treatment, and oily wastewater treatment. However, further exploration and improvement of ENMs are necessary, with a focus on enhancing their mechanical properties and minimizing preparation costs.

In summary, the MD technology has emerged as a highly competitive solution for treating complex hypersaline wastewater, recovering valuable salts, and obtaining distilled freshwater from wastewater. Ongoing efforts by scientists and engineers to develop super liquid-repellent membranes based on electrospinning for MD offer great promise for achieving large-scale applications of this technology in the near future.

Acknowledgements

This work was supported by Nankai University and Cangzhou Bohai New Area Institute of Green Chemical Engineering Fund (20220142), National One Thousand Talents Foreign Experts Program of the Ministry of Science and Technology of China and Tianjin Government (040-BE044741, 040-C021801601).

Appendices and nomenclature

RO	reverse osmosis
MD	membrane distillation
FO	forward osmosis
PVDF	polyvinylidene fluoride
PTFE	polytetrafluoroethylene
DMF	N,N-dimethylformamide
SDS	sodium dodecyl sulfonate
DTAB	dodecyltrimethylammonium bromide
Tween-20	polyoxyethylenesorbitan monolaurate
NIPS	non-solvent induced phase separation
TIPS	thermally induced phase separation
ENMs	electrospun nanofibrous membranes

NMP	N-methylpyrrolidone
CAH	contact angle hysteresis
SA	sliding angle
PFOTES	perfluorooctyltriethoxysilane
CVD	chemical vapor deposition
FAS	fluoroalkylsilane
SiO_2	silicon dioxide
SiNPs	silicon nanoparticle
UOCA	underwater oil contact angle
TiO_2	titanium dioxide
AgNPs	silver nanoparticles
CTS	chitosan
PFO	perfluorooctanoate
WCA	water contact angle
PAN	polyacrylonitrile
PDA	polydopamine
PEI	polyethyleneimine
EDA	ethylenediamine
CNT	carbon nanotubes

Author details

Xiaocheng Zhang[1,2], Yuan Liao[1,2*] and Abdul Ghani Razaqpur[1*]

1 Sino-Canadian Joint R&D Center for Water and Environmental Safety, College of Environmental Science and Engineering, Nankai University, Tianjin, P.R. China

2 Nankai University and Cangzhou Bohai New Area Institute of Green Chemical Engineering, Cangzhou, P.R. China

*Address all correspondence to: liaoyuan@nankai.edu.cn; razaqpu@mcmaster.ca

References

[1] Oki T, Kanae S. Global hydrological cycles and world water resources. Science. 2006;**313**:1068-1072. DOI: 10.1126/science.1128845

[2] Daer S, Kharraz J, Giwa A, Hasan SW. Recent applications of nanomaterials in water desalination: A critical review and future opportunities. Desalination. 2015; **367**:37-48. DOI: 10.1016/j.desal.2015.03.030

[3] Nair M, Kumar D. Water desalination and challenges: The Middle East perspective: A review. Desalination and Water Treatment. 2013;**51**:2030-2040. DOI: 10.1080/19443994.2013.734483

[4] Zhang X, Liao X, Shi M, Liao Y, Razaqpur AG, You X. Guide to rational membrane selection for oily wastewater treatment by membrane distillation. Desalination. 2023;**549**:116323. DOI: 10.1016/j.desal.2022.116323

[5] Liao X, Wang Y, Liao Y, You X, Yao L, Razaqpur AG. Effects of different surfactant properties on anti-wetting behaviours of an omniphobic membrane in membrane distillation. Journal of Membrane Science. 2021;**634**:119433. DOI: 10.1016/j.memsci.2021.119433

[6] Cui J, Li F, Wang Y, Zhang Q, Ma W, Huang C. Electrospun nanofiber membranes for wastewater treatment applications. Separation and Purification Technology. 2020;**250**:117116. DOI: 10.1016/j.seppur.2020.117116

[7] Smolders K, Franken ACM. Terminology for membrane distillation. Desalination. 1989;**72**:249-262. DOI: 10.1016/0011-9164(89)80010-4

[8] Aijaz MO, Karim MR, Othman MHD, Samad UA. Anti-fouling/wetting electrospun nanofibrous membranes for membrane distillation desalination: A comprehensive review. Desalination. 2023;**553**:116475. DOI: 10.1016/j.desal.2023.116475

[9] Lalia BS, Kochkodan V, Hashaikeh R, Hilal N. A review on membrane fabrication: Structure, properties and performance relationship. Desalination. 2013;**326**:77-95. DOI: 10.1016/j.desal.2013.06.016

[10] Liu F, Hashim NA, Liu Y, Abed MRM, Li K. Progress in the production and modification of PVDF membranes. Journal of Membrane Science. 2011;**375**:1-27. DOI: 10.1016/j.memsci.2011.03.014

[11] Yeszhanov AB, Korolkov IV, Dosmagambetova SS, Zdorovets MV, Güven O. Recent progress in the membrane distillation and impact of track-etched membranes. Polymers. 2021;**13**(15):2520. DOI: 10.3390/polym13152520

[12] Francis L, Ahmed FE, Hilal N. Electrospun membranes for membrane distillation: The state of play and recent advances. Desalination. 2022;**526**:115511. DOI: 10.1016/j.desal.2021.115511

[13] He JH, Liu Y, Xu L. Apparatus for preparing electrospun nanofibres: A comparative review. Materials Science and Technology. 2010;**26**:1275-1287. DOI: 10.1179/026708310X12798718274430

[14] Liao Y, Loh C-H, Tian M, Wang R, Fane AG. Progress in electrospun polymeric nanofibrous membranes for water treatment: Fabrication, modification and applications. Progress in Polymer Science. 2018;**77**:69-94. DOI: 10.1016/j.progpolymsci.2017.10.003

[15] Subrahmanya TM, Arshad AB, Lin PT, Widakdo J, Makari HK, Austria HFM, et al. A review of recent progress in polymeric electrospun nanofiber membranes in addressing safe water global issues. RSC Advances. 2021; **11**:9638-9663. DOI: 10.1039/D1RA00060H

[16] Agarwal S, Greiner A, Wendorff JH. Functional materials by electrospinning of polymers. Progress in Polymer Science. 2013;**38**:963-991. DOI: 10.1016/j.progpolymsci.2013.02.001

[17] Sun B, Long YZ, Zhang HD, Li MM, Duvail JL, Jiang XY, et al. Advances in three-dimensional nanofibrous macrostructures via electrospinning. Progress in Polymer Science. 2014;**39**: 862-890. DOI: 10.1016/j.progpoly msci.2013.06.002

[18] Darrell HR, Iksoo C. Nanometre diameter fibres of polymer, produced by electrospinning. Nanotechnology. 1996; **7**:216. DOI: 10.1088/0957-4484/7/3/009

[19] Sill TJ, von Recum HA. Electrospinning: Applications in drug delivery and tissue engineering. Biomaterials. 2008;**29**:1989-2006. DOI: 10.1016/j.biomaterials.2008.01.011

[20] Zong X, Kim K, Fang D, Ran S, Hsiao BS, Chu B. Structure and process relationship of electrospun bioabsorbable nanofiber membranes. Polymer. 2002;**43**:4403-4412. DOI: 10.1016/S0032-3861(02)00275-6

[21] Yang S, Wang X, Ding B, Yu J, Qian J, Sun G. Controllable fabrication of soap-bubble-like structured polyacrylic acid nano-nets via electro-netting. Nanoscale. 2011;**3**:564-568. DOI: 10.1039/C0NR00730G

[22] Hu J, Wang X, Ding B, Lin J, Yu J, Sun G. One-step electro-spinning/netting technique for controllably preparing polyurethane nano-fiber/net. Macromolecular Rapid Communications. 2011;**32**:1729-1734. DOI: 10.1002/marc.201100343

[23] Han J, Xiong L, Jiang X, Yuan X, Zhao Y, Yang D. Bio-functional electrospun nanomaterials: From topology design to biological applications. Progress in Polymer Science. 2019;**91**:1-28. DOI: 10.1016/j.progpolymsci.2019.02.006

[24] Himma N, Prasetya N, Anisah S, Wenten IG. Superhydrophobic membrane: Progress in preparation and its separation properties. Reviews in Chemical Engineering. 2019;**35**(2): 211-238. DOI: 10.1515/revce-2017-0030

[25] Barthlott W, Neinhuis C. Purity of the sacred lotus, or escape from contamination in biological surfaces. Planta. 1997;**202**:1-8. DOI: 10.1007/s004250050096

[26] Teisala H, Tuominen M, Kuusipalo J. Superhydrophobic coatings on cellulose-based materials: Fabrication, properties, and applications. Advanced Materials Interfaces. 2014;**1**:1300026. DOI: 10.1002/admi.201300026

[27] Liu K, Tian Y, Jiang L. Bio-inspired superoleophobic and smart materials: Design, fabrication, and application. Progress in Materials Science. 2013;**58**: 503-564. DOI: 10.1016/j.pmatsci.2012.11.001

[28] Liao X, Goh K, Liao Y, Wang R, Razaqpur AG. Bio-inspired super liquid-repellent membranes for membrane distillation: Mechanisms, fabrications and applications. Advances in Colloid and Interface Science. 2021;**297**:102547. DOI: 10.1016/j.cis.2021.102547

[29] Liao Y, Wang R, Fane AG. Engineering superhydrophobic surface

on poly (vinylidene fluoride) nanofiber membranes for direct contact membrane distillation. Journal of Membrane Science. 2013;**440**:77-87. DOI: 10.1016/j.memsci.2013.04.006

[30] Long M, Ma Y, Yang C, Zhang R, Jiang Z. Superwetting membranes: From controllable constructions to efficient separations. Journal of Materials Chemistry A. 2021;**9**:1395-1417. DOI: 10.1039/D0TA10280F

[31] Chen Y, Kim H. Preparation of superhydrophobic membranes by electrospinning of fluorinated silane functionalized poly(vinylidene fluoride). Applied Surface Science. 2009; **255**:7073-7077. DOI: 10.1016/j.apsusc.2009.03.043

[32] Dong Z-Q, Ma X-H, Xu Z-L, You W-T, Li F-B. Superhydrophobic PVDF–PTFE electrospun nanofibrous membranes for desalination by vacuum membrane distillation. Desalination. 2014;**347**:175-183. DOI: 10.1016/j.desal.2014.05.015

[33] Guo F, Servi A, Liu A, Gleason KK, Rutledge GC. Desalination by membrane distillation using electrospun polyamide fiber membranes with surface fluorination by chemical vapor deposition. ACS Applied Materials & Interfaces. 2015;7:8225-8232. DOI: 10.1021/acsami.5b01197

[34] Dong Z-Q, Wang B-J, Ma X-H, Wei Y-M, Xu Z-L. FAS grafted electrospun poly(vinyl alcohol) nanofiber membranes with robust superhydrophobicity for membrane distillation. ACS Applied Materials & Interfaces. 2015;7:22652-22659. DOI: 10.1021/acsami.5b07454

[35] Liao Y, Loh C-H, Wang R, Fane AG. Electrospun superhydrophobic membranes with unique structures for membrane distillation. ACS Applied Materials & Interfaces. 2014;**6**: 16035-16048. DOI: 10.1021/am503968n

[36] Seyed Shahabadi SM, Rabiee H, Seyedi SM, Mokhtare A, Brant JA. Superhydrophobic dual layer functionalized titanium dioxide/polyvinylidene fluoride-co-hexafluoropropylene (TiO2/PH) nanofibrous membrane for high flux membrane distillation. Journal of Membrane Science. 2017;**537**:140-150. DOI: 10.1016/j.memsci.2017.05.039

[37] Su C, Chang J, Tang K, Gao F, Li Y, Cao H. Novel three-dimensional superhydrophobic and strength-enhanced electrospun membranes for long-term membrane distillation. Separation and Purification Technology. 2017;**178**:279-287. DOI: 10.1016/j.seppur.2017.01.050

[38] Chul Woo Y, Chen Y, Tijing LD, Phuntsho S, He T, Choi J-S, et al. CF4 plasma-modified omniphobic electrospun nanofiber membrane for produced water brine treatment by membrane distillation. Journal of Membrane Science. 2017;**529**:234-242. DOI: 10.1016/j.memsci.2017.01.063

[39] Chen H, Mao Y, Mo B, Pan Y, Xu R, Ji W, et al. Plasma-assisted facile fabrication of omniphobic graphene oxide membrane with anti-wetting property for membrane distillation. Journal of Membrane Science. 2023;**668**: 121207. DOI: 10.1016/j.memsci.2022.121207

[40] Lee J, Boo C, Ryu W-H, Taylor AD, Elimelech M. Development of omniphobic desalination membranes using a charged electrospun nanofiber scaffold. ACS Applied Materials & Interfaces. 2016;**8**:11154-11161. DOI: 10.1021/acsami.6b02419

[41] Lu C, Su C, Cao H, Ma X, Duan F, Chang J, et al. F-POSS based omniphobic membrane for robust membrane distillation. Materials Letters. 2018;**228**: 85-88. DOI: 10.1016/j.matlet.2018.05.126

[42] Lu C, Su C, Cao H, Horseman T, Duan F, Li Y. Nanoparticle-free and self-healing amphiphobic membrane for anti-surfactant-wetting membrane distillation. Journal of Environmental Sciences. 2021;**100**:298-305. DOI: 10.1016/j.jes.2020.04.044

[43] Huang Y-X, Wang Z, Hou D, Lin S. Coaxially electrospun super-amphiphobic silica-based membrane for anti-surfactant-wetting membrane distillation. Journal of Membrane Science. 2017;**531**:122-128. DOI: 10.1016/j.memsci.2017.02.044

[44] Zarghami S, Mohammadi T, Sadrzadeh M, Van der Bruggen B. Superhydrophilic and underwater superoleophobic membranes-a review of synthesis methods. Progress in Polymer Science. 2019;**98**:101166. DOI: 10.1016/j.progpolymsci.2019.101166

[45] Huang Y-X, Wang Z, Jin J, Lin S. Novel Janus membrane for membrane distillation with simultaneous fouling and wetting resistance. Environmental Science & Technology. 2017;**51**: 13304-13310. DOI: 10.1021/acs.est.7b02848

[46] Makanjuola O, Ahmed F, Janajreh I, Hashaikeh R. Development of a dual-layered PVDF-HFP/cellulose membrane with dual wettability for desalination of oily wastewater. Journal of Membrane Science. 2019;**570-571**:418-426. DOI: 10.1016/j.memsci.2018.10.028

[47] Afsari M, Shon HK, Tijing LD. Janus membranes for membrane distillation: Recent advances and challenges. Advances in Colloid and Interface Science. 2021;**289**:102362. DOI: 10.1016/j.cis.2021.102362

[48] Tang M, Hou D, Ding C, Wang K, Wang D, Wang J. Anti-oil-fouling hydrophobic-superoleophobic composite membranes for robust membrane distillation performance. Science of the Total Environment. 2019;**696**:133883. DOI: 10.1016/j.scitotenv.2019.133883

[49] Wang K, Hou D, Wang J, Wang Z, Tian B, Liang P. Hydrophilic surface coating on hydrophobic PTFE membrane for robust anti-oil-fouling membrane distillation. Applied Surface Science. 2018;**450**:57-65. DOI: 10.1016/j.apsusc.2018.04.180

[50] Wang K, Hou D, Qi P, Li K, Yuan Z, Wang J. Development of a composite membrane with underwater-oleophobic fibrous surface for robust anti-oil-fouling membrane distillation. Journal of Colloid and Interface Science. 2019;**537**: 375-383. DOI: 10.1016/j.jcis.2018.11.040

[51] Wang Z, Hou D, Lin S. Composite membrane with underwater-oleophobic surface for anti-oil-fouling membrane distillation. Environmental Science & Technology. 2016;**50**:3866-3874. DOI: 10.1021/acs.est.5b05976

[52] Zhang J, Song Z, Li B, Wang Q, Wang S. Fabrication and characterization of superhydrophobic poly (vinylidene fluoride) membrane for direct contact membrane distillation. Desalination. 2013;**324**:1-9. DOI: 10.1016/j.desal.2013.05.018

[53] An AK, Guo J, Lee E-J, Jeong S, Zhao Y, Wang Z, et al. PDMS/PVDF hybrid electrospun membrane with superhydrophobic property and drop impact dynamics for dyeing wastewater treatment using membrane distillation. Journal of Membrane Science. 2017;**525**:

57-67. DOI: 10.1016/j.memsci.2016.10.028

[54] Attia H, Alexander S, Wright CJ, Hilal N. Superhydrophobic electrospun membrane for heavy metals removal by air gap membrane distillation (AGMD). Desalination. 2017;**420**:318-329. DOI: 10.1016/j.desal.2017.07.022

[55] Attia H, Johnson DJ, Wright CJ, Hilal N. Robust superhydrophobic electrospun membrane fabricated by combination of electrospinning and electrospraying techniques for air gap membrane distillation. Desalination. 2018;**446**:70-82. DOI: 10.1016/j.desal.2018.09.001

[56] Xie S, Pang Z, Hou C, Wong NH, Sunarso J, Peng Y. One-step preparation of omniphobic membrane with concurrent anti-scaling and anti-wetting properties for membrane distillation. Journal of Membrane Science. 2022;**660**:120846. DOI: 10.1016/j.memsci.2022.120846

[57] Hou D, Ding C, Fu C, Wang D, Zhao C, Wang J. Electrospun nanofibrous omniphobic membrane for anti-surfactant-wetting membrane distillation desalination. Desalination. 2019;**468**:114068. DOI: 10.1016/j.desal.2019.07.008

[58] Li J, Ren L-F, Huang M, Yang J, Shao J, He Y. Facile preparation of omniphobic PDTS-ZnO-PVDF membrane with excellent anti-wetting property in direct contact membrane distillation (DCMD). Journal of Membrane Science. 2022;**650**:120404. DOI: 10.1016/j.memsci.2022.120404

[59] Boo C, Lee J, Elimelech M. Omniphobic polyvinylidene fluoride (PVDF) membrane for desalination of shale gas produced water by membrane distillation. Environmental Science & Technology. 2016;**50**:12275-12282. DOI: 10.1021/acs.est.6b03882

[60] Chen Y, Lu K-J, Gai W, Chung T-S. Nanofiltration-inspired Janus membranes with simultaneous wetting and fouling resistance for membrane distillation. Environmental Science & Technology. 2021;**55**:7654-7664. DOI: 10.1021/acs.est.1c01269

Section 2

Applications of Electrospinning

Chapter 4

Electronic and Magnetic Contribution for CuO and CuO Nanofibers Doped with Mn at 3.0%

Manuel F. Piñón-Espitia, Guillermo M. Herrera-Pérez and Matha T. Ochoa-Lara

Abstract

The copper (II) oxide nanofibers (NFs) synthesized with the electrospinning method showed a necklace-like morphology and nanometric size. The use of the XPS (X-ray Photoelectronic Spectroscopy) technique allowed the analysis of the Cu 2p and O 1s orbitals showing a Cu_xO type stoichiometry (x = 1, 2, 3), in turn, the UPS (Ultraviolet Photoelectronic Spectroscopy) region determined the conduction state associated to the dielectric function. These data are compared with the EELS technique. The NFs have presented a behavior with double magnetic phase associated to the non-stoichiometry and oxygen vacancies, and the non-presence of the AFM phase due to the increase of the vacancies. In addition, their electronic and magnetic structure reveal spin-orbit related changes shown in the Cu 2p spectra. The results showed in the conduction band holes and the Cu 2p and O 1s orbitals.

Keywords: NFs, Cu_xO, XPS, UPS, EELS, AFM, electrospinning, Cu *2p*, O *1s*

1. Introduction

The electrospinning technique has been widely used in recent years for the synthesis of nanomaterials, especially micro and nanometer scale fibers for a wide range of applications in areas such as biotechnology, spintronics and electronics. The technique is characterized by being versatile and easy to assemble, allowing the processing of a wide variety of polymers, integrating in recent years ceramics, semiconductors and dielectrics [1]. The technique consists of using polar solutions (PVP or PVA) in which acetates are dissolved, once mixed, the solutions are placed in a syringe connected to a hose and needle, the latter placed at a certain distance to be deposited on an aluminum plate. To this process an electric field is applied between (5 to 20 kV), the process is diversified in terms of uses, time, synthesis process.

The CuO study is a p-type semiconductor with a monoclinic crystal structure, with a C_{2h}^{6} (C2/c space group. The copper atom is coordinated to four coplanar oxygen atoms located at the corner of a rectangular parallelogram and the oxygen atom is coordinated by four copper atoms located at the corner of a distorted tetrahedron. These chains traverse the [110] and $[1\bar{1}0]$ directions, respectively [2].

The crystallographic structure is octahedral. The molecule has 4 Cu atoms in the Wickoff position 4c (1/4,1/4,0) and 4 O atoms in the Wickoff position 4e (0, y, 1/4) where y = 0.4184 [3]. The lattice parameters in the unit cell are a = 4.6837 Å, b = 3.4226 Å, c = 5.1288 Å, the angle β = 99.45° between a and c.

In current reports Cu shows three oxidation states Cu^{1+}, Cu^{2+} y Cu^{3+} [4–6] and excess holes in its valence band structure (VC) which are associated with the presence of Cu^{3+} microtraces [7]. The copper oxide faces V_O due to the synthesis process the crystalline structure undergoes cation exchanges. This originates changes in its chemical coordination octahedral (O_h) or tetrahedral (D_{4h}) generating a superexchange thus giving the magnetic origin [8–11].

In recent studies the oxide has been highlighted by various nanostructures (nanosheets, nanoflowers, nanoparticles, nanorods, etc.) [7, 8, 12–15]. The study of the surface of these nanostructures has shown relevance through X-ray Photoelectron Spectroscopy (XPS) showing a stoichiometric quantification approach and identification of cations/anions. The analysis of CuO has focused on the Cu 2*p* and O 1*s* orbitals, due to its physical characteristics (hybridization and high energy edges). In addition, it is a transition metal because it has incomplete *d* level, therefore, its study is focused on showing its configuration ($1s^2\ 2s^2\ 2p^2\ 3s^2\ 3p^6\ 3d^9\ 4s^2$) and the acceptance of electrons from the O^{2-} ion, this phenomenon allows charge transfer. However, this is not orange blossom, because a chemical coordination O_h and D_{4h} is governed. These coordination's show distortions caused by the packing of ions of different sizes generating the Jahn-Teller effect. Other authors have shown that the 3*d* orbital shows perturbations due to the possibility of displacements due to bond distortion, clarifying that it is not due to the generation of V_O [12–16]. Some authors have focused on studying the Cu 3*d* orbital because of its proximity to the VB [17]. Some authors have shown modifications in their forbidden gap (E_g) because of quantum effects forbidden in the continuum. These changes are due to the quantum confinement of the electronic states which are modified by states not yet clarified in the literature because on the one hand the crystal structures are modified due to the nanometric dimensions and their resonance from the VB to the CB [9–11, 18–24].

XPS is used to understand the oxidation behavior of metals. The study of CuO NFs has led to the elucidation of the initial excitation processes [5, 6, 24–26].

The analysis of the Cu $2p_{3/2}$ and Cu $2p_{1/2}$ doublets shows changes in the electrical behavior. In addition, the association of the O 1*s* orbital, which is quantified to obtain the stoichiometry of the oxide, has been added to this study. The spectroscopists suggest to calculate these by the block method which provides higher accuracy ±4.0% error [27]. On the other hand, the background of the spectra is important to associate chemical species, this quantification is relevant due to the resonance of electrons in the valence band [28].

The 2*p* photoemission spectra of the transition metals in the first row are distinguished by being highly complex, showing a large and complex background, peak asymmetries and a broad multiplet structure. However, the Cu 2*p* spectrum is particularly peculiar, even among them. We will discuss these peculiarities and how the associated difficulties can be overcome by using the Shirley - Vegh - Salvi - Castle (SVSC) based fitting methods under the active approach, as well as compare with the Tougaard contribution, in order to know the electronic contribution associated with the chemical species in the CuO [5, 29, 30].

The study for the identification of the chemical state and its quantification by XPS, using the NIST database [31] and Electron Effective Attenuation Length (EAL), generally provide the necessary information for the identification of metal oxides.

Investigations of multiplet structure calculations suggest that transition metals exhibit charge transfer which can complicate the intensity of the spectral multiplet peaks. Analysis of 2p spectra can be complicated to distinguish the ion structure and charge effect of neighboring bonded ions as both the oxidation state and the multiplet splitting can be affected by this perturbation. These phenomena have been observed principally in copper oxide [5, 32, 33].

The multiplet structure in recent years has been analyzed using CTM4XAS (Charge Trasfer Multiplet) software. The software solves quantum data through density functional theory (DFT) and Hartree-Fock function to calculate the spin-orbit entanglement.

DFT (Density Functional Theory) and the Hartree-Fock function to calculate the spin-orbit entanglement. The calculation is used to determine the final energy of the orbital to be treated using three primitives: (a) atomic multiplet, (b) crystal field and (c) charge transfer. Employing the Schrodinger equation, the minimum energy, and e-e interactions and spin-orbit interaction are determined by providing information of the electron location employing the atom symmetry and charge fluctuation in the energy states $(3d^n \rightarrow 3d^{n+1}L)$, considering L as a hole, respectively. The multiplet structure is important in the study of XPS due to the contribution of the VB electrons and the core-hole (generated from the photoelectric effect). Moreover, in transition metals the 3d ($L_{3,2}$) edge is the most important due to the unpaired electron, which provides electronic information [30, 31]. Stavitski and de Groot [32], point out that one of the first features of XPS is the absorption of electrons in the BV and the core-hole effect, because of this the $L_{3,2}$ edge is important due to the $2p^6 3d^n \rightarrow 2p^5 3d^{n+1}$ transition, due to this absorption the multiplet effect is affected in the deeper $2p$, $3s$ or $3p$ levels [30, 32].

The calculation process using CTM4XAS software employs empirical data: a reduction parameter of the Slater integral, which refers to the final configuration of the calculated electronic structure. The crystal field (10 Dq) would indicate the separation of e_g and t_{2g}. Δ (charge transfer energy) refers to the average energy for the $3d$ orbital, Udd (the core hole potential) is the energy transferred to a hole (unoccupied state) and Upd (Hubbard) is a potential related to the charge fluctuation in the 3d orbital. In addition, the Lorentzian and Gaussian values taken from some database are fixed [32].

Furthermore, the study of the multiplet structure of CuO has not been generated in recent years unlike the undersigned of this manuscript [26, 33], therefore, authors Okada and Kotani to extend and determine the data obtained [25]. These authors studied three Cu oxides by approximation calculations using a Hamiltonian proposing the energy of the free ion (Cu) and the gap (electronic repulsion due to its neighboring O 2p atom) in the BV, binding the energy level Cu $2p^6$ $3d^9$ $4s^2$ and its final state $3d^8$ L (where L denotes the gap and the relationship with O $2p$). This spectrum has been discussed by several authors due to the importance of the Columbian interaction, spin-orbit entanglement and the screening shielding effect [25].

The EELS technique evaluates the energy loss of inelastic scattered electrons passing through the sample in the transmission electron microscope (TEM). The valence electron energy loss study (VEELS) focuses on the lower energy loss spectrum, typically <50 eV, this region shows the valence and conduction band characteristics due to electron beam interaction. These are appreciated using scanning transmission electron Microscopy (STEM) Park and Yang [34]. Egerton mentions that the use of STEM provides nanometer resolution spectra, thus allowing surface and crystal structure information to be obtained [35]. The author Herrera-Pérez et al. presents the VEELS technique as a approach to obtain Eg characteristics, electronic and electrical properties through the cole-cole diagram (the graph is obtained from the Im vs. Re function), from the Single Scattering Distribution (SSD) [36]. This analysis is important to determine the resolution of the Zero Loss Peak (ZLP) from the method proposed by Park and Yang [34]. In addition to the above, obtaining the modern spectra and analysis proposed by Eljarrat et al. from the Kramers-Kronig analysis (KKA) and the elimination of the relativistic effects (Cerenkov effect) [37]. Herrera-Pérez et al. shows in their optoelectronic studies the elimination of this effect, as well as convenient methods to calculate the complex dielectric function (CDF) of the materials [36]. In recent years, the electronic study has been complemented by joint density of states (JDOS) convolution showing the electrons present in the conduction band [38].

On the other hand, the theoretical calculations obtained by means of density functional theory (DFT) show similarities to the experimental ones by EELS and XPS, techniques that are exposed in this research, but how to approximate it, for years it has been tried to solve the Schrodinger equation, Born-Oppenheimer in the early 30's proposed an equation to model multielectronic systems (a many-body wave solution) with N variables. The first-principles or ab initio method is based solely on the laws of physics, without reference to empirical parameters. Kohn-Sham to give solution to this uncertainty treats the set of N particles as electron density, broadens the potential energy with a potential that has been neglected to obtain the non-interacting parts (VXC Exchange-correlation potential) in the Hamiltonian (the Hamiltonian supplements ψ in the Schrödinger equation; the latter has no physical meaning); relative to the term shows that the motions of the electrons are correlated with each other. Software (CASTEP -Cambridge Serial Total Energy Package-, Wien2K, VASP) have implemented for such theory two approximations: Local Density Approximation (LDA) and Generalized Gradient Approximation (GGA), which perform estimates considering variables for Eg's energy solution, density of states (DOS). However, the calculations have shown estimates that do not agree with those measured experimentally, especially for the d and f orbitals. In the last two decades Hubbard - U effects implemented in DFT have been shown, described as DFT + U in strongly correlated systems that approximate the d and f systems to the expected ones. What does the Hubbar value in the approximation mean, it is a value that serves to describe the electron-electron interaction without shielding (e.g., in the CTM4XAS software it uses the Slater integral to consider such interaction to estimate the type of chemical bond), this interaction is important to determine the magnetic moment and E_g.

The aim of this work is to determine by modern fits to XPS and EELS spectroscopies the Cu 2*p*, O 1*s* electronic states and the interband transition states. Furthermore, to calculate the stoichiometry of the A-NFs and the reference (Sigma Aldrich). The EELS section will show a comparative study of CuO and CuO doped with 3.0% Mn to know the changes associated with Mn compared to that calculated with CASTEP 7.0 software. Finally, the magnetic study will show the change of magnetization with respect to temperature and the association of the magnetic field with respect to the shape of the NFs.

2. Experimental procedure

For the synthesis of polymeric fibers by the electrospinning method, the following reagents were used:

1. PVA (High purity, 130,000 Da, Sigma Aldrich)

2. Tri-distilled water (1.1 μohms/cm, J. T. Baker)

3. Copper Acetate II (99.6% of purity, Sigma Aldrich)

The 8.0% PVA polymer solution was made by weighing 8.0 g of PVA in tri-distilled water at 100°C with constant agitation at 100 rpm for 24 hours.

For the polymeric solution of copper acetate II, 1.0 g of it was weighed and added over 2 ml of tri-distilled water at 50°C and left stirring for 2 hours. Finally, 10 g of PVA was added and the whole solution was stirred for 24 hours.

Finally, the polymer solution was placed in a 10 ml syringe connected to a hose for dosing by means of a number 22 needle. The parameters used in the pump were 0.3 ml/hr. and the distance from the needle to the collection plate was 20 cm.

2.1 Characterizations and computational methods

2.1.1 Scanning electron microscopy (SEM)

The sample used was a piece of aluminum with an area of 1 cm^2, taken from the collector sheet, and used as an electrode for electrospinning. The collected images were taken from the SEM Hitachi SU3500 equipment used to verify that the material was obtained in the form of fibers.

2.1.2 TGA-DSC analysis

The TGA-DSC analysis aimed to show the chemical reactions with their weight distribution in relation to the heat of the reaction (entropy).

A simultaneous thermal thermogravimetric difference test (TGA-DSC) was performed using the SDT-TGA Q600 equipment of TA instruments, through the air and a temperature ramp of 10°C/min, with a range of 25–1000°C, placing 0.25 mg of the synthesized material in a sample holder of the equipment, to determine the calcination temperature of the study compound.

2.1.3 Structure characterization and XRD analysis

Structure characterization was performed by X-ray diffraction (XRD) using an X'Pert Pro diffractometer equipped with an X'Celerator detector. Diffraction patterns were taken in an interval $2\theta = 30°–65°$ using a step size of 0.05 s^{-1} with a $Cu - K\alpha$ radiation source $\lambda = 1.5418$ Å.

From the lattice parameters obtained from the XRD characterization, these spectra were used for Rietveld refinement using a Pseudo-Voight peak shape (least squares method) in Fullprof suite software version 2022 [39]. A CIF structure was obtained from the refinement, which was used to determine the 2-D and 3-D electron density using VESTA software (free version provide by VESTA) [40]. Especially this analysis took into account the thermal factors of the crystal structure of Arbrik et al. to optimize the calculated versus experimental spectrum using the above-mentioned software [3].

2.1.4 High resolution-transmission electron microscopy (HR-TEM)

The synthesized samples were dispersed at 10 mgL^{-1} isopropanol for 20 min, then deposited on a 3 mm diameter nickel grid with carbon membrane.

The BF-HR-TEM micrographs were obtained in the JEOL-JEM-2200FS equipment operated at 200 kV. In addition, JImage software was used to determine the average particle size distribution by entering the micrographs and calibrating each one. In this way a histogram was constructed using a Log Normal function [26].

2.1.5 X-Ray photoelectronic spectroscopy (XPS)

The XPS study was obtained using the following parameters: scan energy 10 eV, resolution 0.1 eV, dwell time 200 msec, 40 scanning scans per spectrum and a 90° angle. The monochromatic source used was an Al Kα ($h\nu = 1486.6$ eV).

The nanofibers were deposited on a graphite tape, adhered to the sample holder of the XPS equipment, introduced into the pre-chamber at a vacuum pressure of 10^{-6} Torr, and finally into the analysis chamber at a pressure of 10^{-10} Torr.

The peak analysis of the spectra was calculated using AAnalyzer® software (DEMO version) by means of Gaussian and Lorentzian functions (block method, which consists of assigning individual values to each peak, that is, it is not cumulative [27]), these areas obtained were used to calculate the stoichiometry of the oxide. In addition, the convolution considered the Shirley model type SVSC (Shirley-Vegh-Salvi-Castle) suggested for high resolution spectra (in this case Cu 2p), also, for the case of O 1s a Slope line was applied. The data were plotted in Origin Pro 8.6 software [26].

To calculate the chemical composition of NFs, a spherical model was used (model suggested by Shard et al. [41] for 1D materials) that considers the electron attenuation effect (λ), in relation to this Bravo-Sánchez et al. [42] and German-Cabrera et al. [29], relate the λ factor for 2D materials (the difference of this method to the one mentioned above is that it does not consider the shape) as the effect of electron interactions with the atoms of the material, thus affecting the binding energy (E_b) [43, 44].

2.1.6 CTM4XAS

The free version of CTM4XAS software was used to elucidate the multiplet structure (possible energy states calculated by the Schrodinger equation) for the Cu 2p spectrum. The calculation procedure consisted of choosing the Cu element with valence 2+, then the XPS 2p option. By choosing XPS, the charge transfer flag was automatically activated. In addition, the stacks options (to plot the multiplet structure) were selected. To accurately determine the theoretical spectrum, a tetrahedral symmetry was chosen as suggested by Okada et al. [25], the spectrum obtained was calibrated at 933.60 eV to compare the experimental (A-NFs) and reference spectra (Sigma Aldrich) [43].

2.1.7 Electron energy loss spectroscopy (EELS)

The study of NFs by EELS spectroscopy was carried out on a Gatan spectrometer (Tridiem 866 ERS), coupled to a Titan G2 60–300 transmission electron microscope. A Wien monochromator was used to correct the electron beam, the operating conditions were carried out at 200 kV, the collection angle was 17 mrad (corresponding to a parallel beam α = 0 mrad), in diffraction mode at 0.05 eV/Channel, a pitch of 5.0 nm and 2.0 s per pixel.

Through the valence electron energy loss spectroscopy (VEELS) method, and using a scanning transmission electron microscope (STEM), the spectra of A-NFs and B-NFs (CuO, doped with 3.0% Mn) were fitted to the center of the zero loss peak (energy calibration) using the asymmetric Pearson VII function to calculate the experimental resolution, this calculation will be carried out through Origin Pro 8.6 software [44].

On the other hand, the data obtained were processed in the DigitalMicrograph software where the measurement noise in the spectrum was subtracted. Also, the dielectric function and the loss function were obtained mainly by means of the Kramers-Kronig analysis. To remove the relativistic effect, they were treated through a difference method working with angles of: β1 = 17.3 mrad and β2 = 9.1 mrad, that is, two spectra taken with different angles were subtracted. This method is used to determine the bandgap [36].

On the other hand, the thickness value of the samples was calculated using the Log-ratio method (absolute, with relative thickness), taken from the author Egerton [35], from the parameters of: electron beam energy E_0 (kV), the effective collection angle (β), and the effective atomic number Z_{eff}. In addition, take into account t = λ/0.75, where λ is an attenuation factor of the electron on the sample (statistical value, this value is taken from a graph proposed by the author Egerton [35]. To obtain the single scattering distribution (SSD), it was analyzed by means of the Fourier function using the following equation:

$$S\,(E) = \frac{I_0 t}{\pi a_0 m_0 v^2}\, Im\left(\frac{-1}{\varepsilon\,(E)}\right) \ln\left(1 + \left(\frac{\beta}{\theta_E}\right)^2\right) \tag{1}$$

where I_0 is the intensity of the zero-loss beam, t is the thickness of the sample, v is the incident electron velocity, β is semi-angle of collection, a_0 is the Bohr radius, m_0 is the rest mass of the electron and θ_E is the characteristic scattering angle for energy loss [34, 45].

To obtain the electron loss function (ELF) was from the SSD employing Kramers-Kroning [35] given by:

$$1 - \frac{1}{n^2} = \frac{2}{\pi}\int Im\left[\frac{-1}{\varepsilon(E)}\right]\frac{dE}{E} = \frac{2}{\pi}\int [scale\_Factor \times SSD]\frac{dE}{E} \tag{2}$$

The scale factor was determined using the refractive index of CuO (2.3), the same as that used for the doped material. Then the real function of the dielectric function, Re[1/ε(E)], experimentally obtained from the loss function, Im [−1/ε(E)], using the Kramers-Kronig transformation.

$$Re\left[\frac{1}{\varepsilon(E)}\right] = 1 - \frac{2}{\pi}P\int_0^{\infty} Im\left[\frac{-1}{\varepsilon(E)}\right]\frac{E' dE'}{E'^2 - E^2} \tag{3}$$

where P denotes the Cauchy integral part (it is a weighting to correct for any imbalance in the sample data). Finally, the complex dielectric function ($\varepsilon = \varepsilon_1 + i\varepsilon_2$), ε_1 and ε_2, was obtained from the relation Re[1/ε(E)] and Im [−1/ε(E)] [34].

To distinguish the inter-band transitions from the imaginary function, according to the equation [35, 36]:

$$J_{CV} = J_{cv1} + i\,J_{cv2} = \frac{m_o^2}{e^2\hbar^2}\frac{E^2}{8\pi^2}i(\varepsilon_2(E) + i\varepsilon_1(E)) \tag{4}$$

where m_0 is the electron mass, e is the Charge and E is the energy, $J_{CV} = \mathrm{Re}[J_{CV}]$ is the joint density function of states, these provide the inter-band transitions obtained from the imaginary part of the dielectric function.

As a comparative method to the complex dielectric function and the joint state function, these were calculated using Materials Studio 7.0 software, through the CASTEP (Cambridge Serial Total Energy Package) program using 10 cores on the Xeon10 server of the CIMAV cluster, Chihuahua, Mexico. This was possible using a P1 crystal structure, and atomic positions of Cu (0.25, 0.25, 0) and O (0, 0.4185, 0.25). The structure was optimized and energized using the GGA (Generalized Gradient Approximation) method and the PBE (Perdew Burke Ernzerhof) functional. The convergence was place at $2{\bullet}10^{-5}eV/atomo$, the force on the atom 0.01 eV/Å, the stress on the atom less than 0.02 GPa, and the maximum atomic displacement no more than $5{\bullet}10^{-4}$Å. The electron exchange of the correlation energy was treated in the GGA framework using PBE, as well as a U potential (7.5 eV) as a correction method in the forbidden band, the energy cutoff of the plane wave basis set was chosen at 500 eV. Directions over the Brillouin zone (BZ) were fixed by the Monkhorst-Pack method with a grid (k-points) of $8 \times 8 \times 8$. In addition, the convergence criterion for the total energy was set with a self-consistent field (SCF) tolerance at $2{\bullet}10^{-6}eV/atomo$. For the calculation as an antiferromagnetic structure without fixed spin orientation. The DOS (density of states) values were taken from the calculations suggested by the author Absike et al. [46].

3. Results and discussion

3.1 SEM analysis

The polymeric fibers obtained by the electrospinning technique were analyzed through SEM, in the secondary electron mode, at magnifications of 5000×. The images are shown in **Figure 1a**, by means of JImage software calculated the average sizes of the fibers (panel b), resulting to be 47.24 nm.

3.2 TGA-DSC analysis

The graphs shown in **Figure 2** present the result of the simultaneous TGA-DSC test. The same figure illustrates the synthesized and calcined material (panels a and b).

This technique was used in the thermal analysis to find the calcination temperature, which after analyzing panel d illustrating the different chemical reactions of decomposition in the polymeric fibers until CuO was obtained, was found after 650°C. Temperatures from 40.5–650°C show the decomposition of the organic and inorganic groups that compose the polymeric fibers such as: alcohol, water, chemically bound water, acetate, and PVA, mainly (see panel d). The analysis was complemented with a DSC that exhibits two types of reactions for the research materials: positive enthalpy

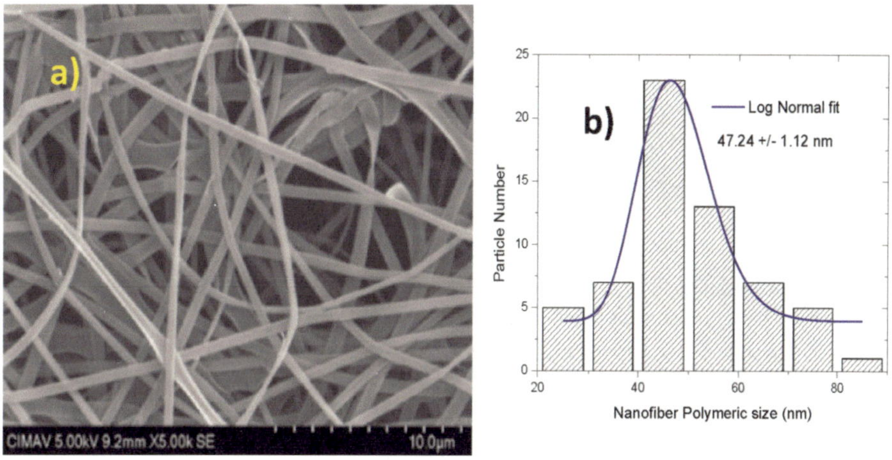

Figure 1.
Polymeric fibers obtained by electrospinning method, CuO (panel a), and average diameter quantified by JImage software shown in panel b.

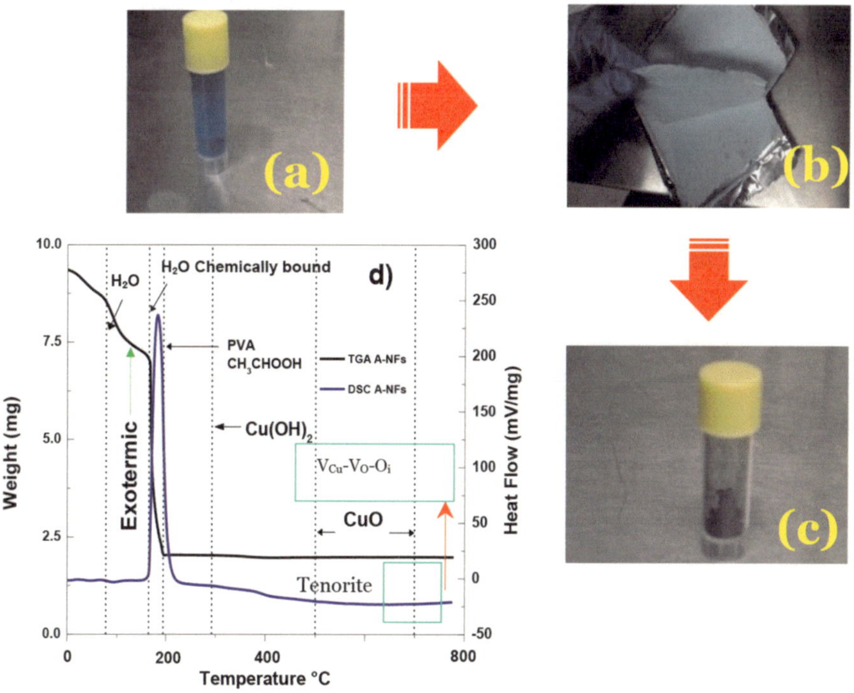

Figure 2.
Panel (a) illustrates an example of the finished polymer solution for use in electrospinning. Panel (b) shows the electrospun polymeric fibers and panel (c) shows the calcined polymeric fibers. Finally, panel (d) illustrates the TGA-DSC analysis showing the decomposition of the precursors of the polymer solution.

(exothermic), related to the release of thermal energy, manifested by the peak at 230 J/g (180°C), and a negative enthalpy (endothermic), related to energy adsorption −25 J/g, at 650°C. The final concentration shows thermal stability for pure CuO, corroborating that, from 650°C, there are no energy and mass changes, respectively.

On the other hand, the authors Kroger et al. [47], Jeong et al. [18], and Piñon-Espitia et al. [26] indicate that the negative enthalpy energy is related to the formation of V_O (oxygen vacancies) from an uncontrolled atmosphere.

Based on the results shown in **Figure 2d**, in an uncontrolled atmosphere, the defects associated with V_O are corroborated by TGA due to a negative enthalpy.

3.3 DRX analysis

Figure 3 shows the XRD obtained from the NFs, indicating that the predominant phase is tenorite (panel a), which was corroborated with PDF 80–1268 (panel b). Relation to the fits by the Rietveld method show agreement with the experimental result (Y_{obs}-Y_{calc}), which is shown in the same figure.

In **Table 1**, the variations related to the Rietveld refinement settings for the XRD spectra are presented. Their variants allow elucidation of their crystal structure with respect to the average crystallite size, lattice parameters, and changes in the O^{-2} anion in the crystal structure.

The parameters a, b, and c of the structure are compared with those of the reference sample of Asbrik [3], showing consistency with the calculated ones. In addition, the Y parameter shows changes due to the presence of oxygen vacancies [40, 48].

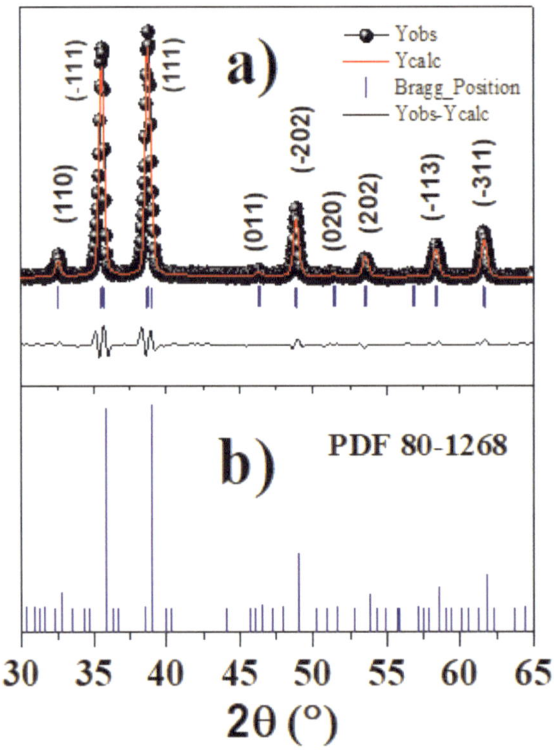

Figure 3.
Experimental XRD data are shown (panel a), which are compared with the PDF (power diffraction file) data shown in panel b.

Structure	a (Å)	b (Å)	c (Å)	V (Å^3)	χ^2	Y (O^{2-}) (Å)
Sigma Aldrich	4.6837	3.4226	5.1288	81.199	—	
Asbrik [3]	4.68370	3.42260	5.12880	81.079	—	0.41840
A-NFs	4.6858	3.4230	5.1298	82.282	1.44	0.41600

Table 1.
The data obtained from the Rietveld refinement for the NFs under study compared with two references (sigma Aldrich and S. Asbrik) are shown.

3.3.1 Electron density in the tenorite structure

According to the XRD results explained above, the structure obtained from the refinement is shown in **Figure 4**. The 3-D structures exhibit the (110) plane. In panel a shows the comparison of the electron density for the A-NFs and the references (Asbrik [3], Cao et al. [49]), indicating charge transfer from oxygen to Cu^{2+}. In relation to this point, this phenomenon is characteristic of MOs as a consequence of thermal treatment, nanometric dimensions, and thermodynamic changes (the oxide formation energy change generally presents negative peaks associated with endothermic absorption) [16, 32, 40, 48, 50, 51]. In addition, the energies exhibited in panel a suggest a covalent bond, increasing due to energy transfer this suggests an increase in the forbidden E_g and a greater number of defects [40, 48].

3.4 TEM analysis

According to the XRD analysis, morphological changes were studied compared to the Sigma Aldrich reference sample.

The bright field (BF) analysis by HR-TEM shows in **Figure 4b** the formation of fibers in the form of a collar, which shows the characteristic planes of the tenorite phase (diffraction pattern attached to the image) compared to the Sigma Aldrich reference (panel b). In addition, an analysis was performed concerning the defects shown in panel c (it was amplified in panel d) and utilizing the edge marked with the yellow arrow in panel c. An EELS analysis (high energy between 400 and 700 eV) was performed to identify oxygen, this data corroborates the TGA-DSC analysis explained in Section 3.2.

The distribution and elemental composition of the NFs, using the EDS of the Hitashi 7700 TEM. Panel a show the pure CuO NFs showing the Cu (yellow) and O (red) components, furthermore, this can be corroborated with the signals obtained from the EDS analysis (see panel b). The distribution of the elements in the mapping NFs showed homogeneity, in a 1:1 ratio (see reference [33]).

3.5 XPS and CTM4XAS analysis

The XPS results are displayed in **Figure 5**, these plots show the spectral differences and spin-orbit coupling energy for the A-NFs and the Reference (Sigma Aldrich). The spectra in panels a and b show the contributions of Cu^{1+}, Cu^{2+}, and Cu^{3+} cations in the main Cu $2p_{3/2}$ and Cu $2p_{1/2}$ peaks. The presence of Cu1+ and Cu2+ cations are bound to oxygen (panel c and d), this can be corroborated by the NBD diffraction patterns in

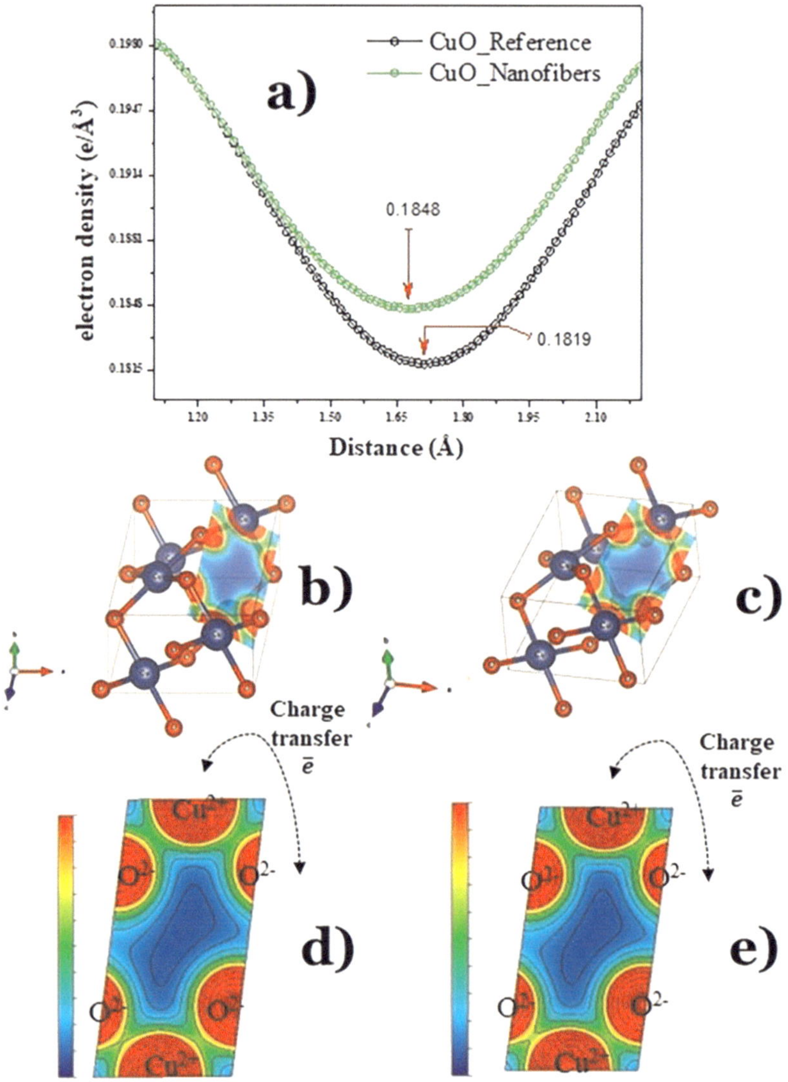

Figure 4.
In panel (a) the comparison of the electron density of S. Asbrik [3] and A-NFs is observed, in panel b and c the 3-D crystal structures of CuO in the (110) plane are shown, respectively. Finally, in panels (d) and (e) the 2-D electron density is plotted for the reference and experimental structure.

Refs [48, 52]. These spectra show a spin-orbit splitting $\vec{J} = \vec{L} + \vec{S}$ at the peaks located at: 19.97 and 19.91 eV, respectively. Furthermore, it can be corroborated that in the left panel of **Figure 3**, the thin films reported by Pauly et al., up on oxidation of the orbitals also present a splitting attributed to the thermal treatment of the material [53].

Three settings indicating three valence states for Cu (1+, 2+ and 3+) have been proposed in the main branches, reported by the authors: Piñon-Espitia, et al., Ochoa-Torres, et al., Pauly, et al., Sarkar, et al., [5, 26, 50, 53]. These authors have also discussed the presence of mixed oxides by the chemical reaction Cu_2O.

On the other hand, these peaks are related to phenomena such as: paramagnetism and electrical conductivity, due to the presence of the Cu^{2+} cation associated to the

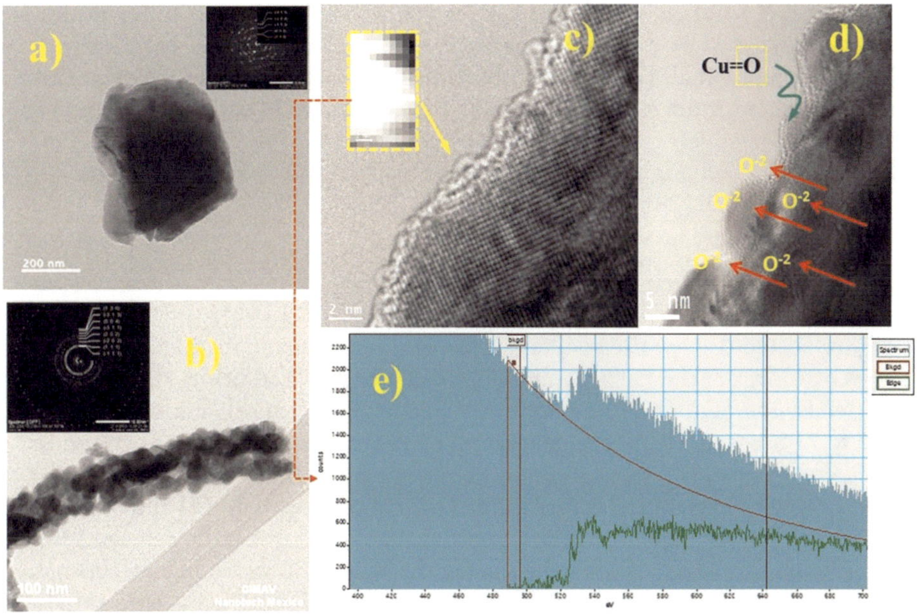

Figure 5.
An analysis of the A-NFs is shown regarding the corroboration of the tenorite phase shown in panel b, compared with the sigma Aldrich reference (panel a), these show the electron diffraction pattern which shows the characteristic planes (−111) and (111). In panel c and d, a surface analysis of the A-NFs due to the defects shown in the edges is exhibited, panel e shows by EELS (high energy edges) the presence of oxygen.

CuO [54, 55]. In **Figure 5** The contributions of the chemical species present in the peaks (Cu^{1+}, Cu^{2+}, Cu^{3+}) given by the areas (see **Table 2**) indicate that the greatest contribution to the conductivity is due to the cations of the A-NFs, which coincides with their geometry that includes a greater area-volume ratio compared to the Reference (Sigma Aldrich).

In the case of Cu-bonded oxygen, the evolution in **Table 2** of sample A and the commercial sample is presented. There is a relationship between the areas between the cation and the anion (oxygen). In the particular case of oxygen, the reactions obtained for this work have been shown in Section 3.2, as well as their derivatives (e.g., enthalpy and entropy). Sample A showed a higher amount of Cu^{1+} as well as a tendency to decrease the amount of the other cations, so according to the author Chusei et al. [56] there is a tendency to higher reactivity because the surface oxygen area increases according to the Cu^{1+} peak, meaning that it increases. While the amount of oxygen vacancy is like the Reference.

Sample	Cu^{1+} 2*p* (eV)	O 1*s* Cu^{1+} (eV)	Cu^{2+} 2*p* (eV)	O 1*s* Cu^{2+} (eV)	Cu^{3+} 2*p* (eV)	Vacancies (eV)
A-NFs	19104.6	4543.56	16063.7	5865.18	15,228	869.60
Reference	2357.15	7766.1	5944.72	10721.00	6214.70	3169.50

Table 2.
Analysis of areas for the adjustments made.

On the other hand, as mentioned in the experimental part, the Shirley of the spectra was analyzed by modifying the block method and the SVSC function proposed by Herrera-Gómez et al. [28], adding the Tougaard function to the analysis. The calculated values for the A-NFs and the Reference were: 0.24 to 0.04 eV^{-1} and 3864 to 2000 eV^{-1}, respectively, associated with recombination of the 2*p* band with the conduction band (see **Figure 5**), which is corroborative with **Figure 3** of author Herrera-Gómez et al. [28].

3.5.1 Analysis of O 1s spectra and Vo contributions

Figure 5c and **d** show the results of the O 1*s* spectrum and its homogeneity in the samples by means of the Background Slope calculation. In addition, this analysis is supported by the TGA-DSC technique (see Section 3.2) that other authors have related to XPS, especially on the generation of vacancies in the tenorite structure [55, 57].

The vacancies have been located in the O 1*s* spectra between 531 and 533 eV and are related to the enthalpy due to the negative reaction suggested by the authors [48, 55] in the formation of the oxide, being the TGA-DSC technique an alternative for the detection of vacancies (they have been identified as S2 in panels c and d). The importance of the generation of V_O is how they influence its physical properties such as magnetic, in addition to the electrical properties of the material indicated by the literature [41, 58–61].

Table 3 shows the percentage variations of Cu^{1+} and Cu^{2+} cations related to the O2- ion, which are compared with the Reference (Sigma Aldrich). The A-NFs have a Cu^{1+}/O^{2-} ratio of 0.92 and 0.72% for the Reference. The percentage corresponding to the Cu^{2+}/O^{2-} ratio is 1.08% in the A-NFs, and 1.38% for the Reference. These non-stoichiometric ratios are associated with structural and surface defects and other phenomena related to the nanometer dimensions of the material (e.g., plasmonic effects) [62, 63].

These percentage ratios show that the A-NFs presented an increase of Cu^{1+} in the crystalline structure, compared with the Reference which has a lower amount and a

Sample	Pico Energía (eV)	$[Cu^{1+}]$%	$[Cu^{2+}]$%	V_O
A-NFs	529.36	0.92		
	531.35		1.08	
	532.83			0.096
	536.06			
Reference	529.83	0.72		
	530.95		1.38	
	531.87			0.653
	532.65			
	533.66			
	535.81			

In addition, the quantification of V_O, as well as the ratio of V_O with respect to the network oxygen.

Table 3.
Percentage of Cu^{1+} and Cu^{2+} spices associated with oxygen ($O_1(S_0)$-$O_2(S_1)$), respectively.

higher amount of Cu^{2+}; in relation to this, the research samples have slightly an excess of Cu^{2+}, this shows that the presence of both cations are important for their electrical, magnetic and optical properties, mainly [12, 54, 61, 64].

3.5.2 Multiplete structure analysis

In this section we show evidence of changes in chemical coordination, binding energy, hybridization, and charge transfer (Δ), related to the nanometer dimensions of the synthesized material compared to the Reference (Sigma Aldrich). Other related changes are: scramming shielding effect, electronic changes in the main $2p_{3/2}$ peak (Cu $2p$ – $L_{3,2}$ edge electrons) and satellites (holes).

Figure 6 shows the final states of the NFs compared to the commercial and calculated material (CTM4XAS). Image a show the set of study spectra compared to the calculated one (spectral shape and multiplet structure). In addition, the panels show the changes mainly in the satellites discussed in the previous section [32, 65, 66].

In **Figure 6a** the Δ according to the obtaining of the materials is exposed, which is seen to be higher for the A-NFs and much lower for the commercial one. According to Okada and Kotani [25] the Δ effects are due to the electron donors (in this case the two oxygens located in the c plane), data that agrees with the percentages presented in **Table 3**, which would show that the A-NFs show higher charge transfer due to the proximity of the oxygen ion to Cu and generating a higher scramming shielding.

The authors showed that the satellites are due to the holes in the oxide, which is observed in panel a that physically exemplifies the effect of the exposed edge related to electrons and holes in the BV. Finally, the leading-edge shifts and hole variation are

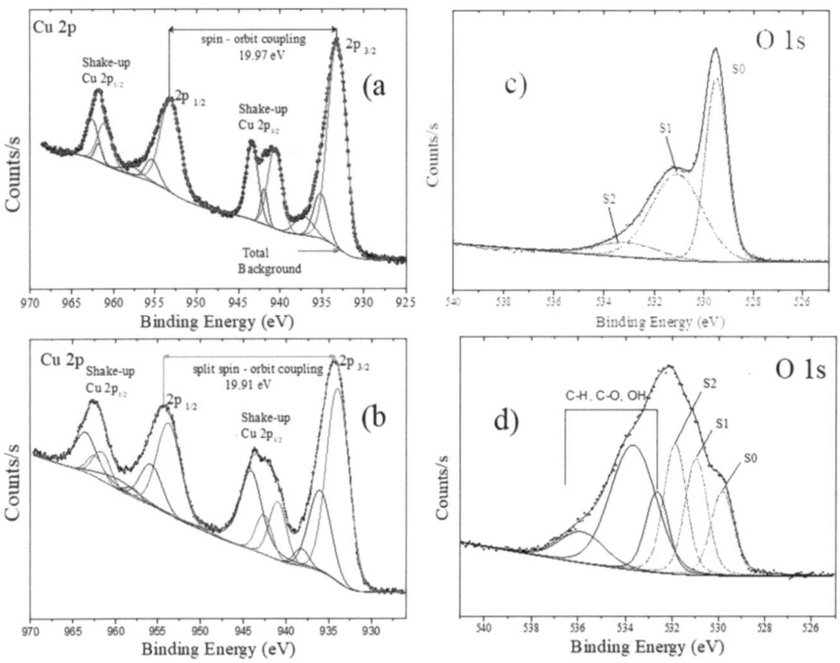

Figure 6.
Panels a and b show the Cu 2p spectra of the A-NFs and the reference (sigma Aldrich) and O 1 s (panels c and d), respectively.

mainly due to the hybridization of the Cu 2*p* and O 1*s* edges, and their shift to higher binding energy, as is the case in the experimental sample exhibiting the screening shielding effect.

About the satellites, it can be appreciated that according to the ions and their coordination, variations were obtained with those shown in **Figure 5** (image a and b), indicating for the experimental ones a D_{4h} type structure (tetrahedral) and the Reference type O_h (octahedral). On the other hand, the screaming shielding effect is directly related to electronic effects related to electron and hole resonance.

3.5.3 Chemical composition analysis based on a spherical model

The chemical composition of NFs has been calculated through the spherical model. This proposal is derived from recent reports by Shard et al., and Cardona, [61, 67]. In **Table 4** shows that A-NFs do not retain stoichiometry.

According to the geometries estimated with the spherical model of Shard [41, 61] the compositions were for A-NFs: $Cu_{1.02}O_{1.16}$, $Cu_{2.22}O_{2.12}$, and for the Reference (Sigma Aldrich): $Cu_{0.97}O_{1.02}$, $Cu_{2.43}O_{2.01}$, with a 4.0% error, respectively. The chemical composition was related to the E_k and to the photoelectric effect applied to the sample (cross section).

3.6 EELS analysis

For this analysis, two samples, A-NFs and B-NFs (CuO doped with 3.0% Mn) were taken for comparison. To corroborate the distribution of Mn and its percentage can be observed in the reference [33].

Figure 7 shows the inelastic contributions of the electrons in the A-NFs and B-NFs, respectively. The resolution of the peaks exhibited in panels a-b was 0.97 eV, as an asymmetric result of the deconvolution. In addition, the loss function (blue line) is shown in both panels, in panel a a a plasmon is exhibited at 22 eV, in panel b 2 plasmon peaks are exhibited, 10.0, 22 eV, while the ~32.0 eV peak represents a $M_{2,3}$ Mn [35]. On the other hand, the analysis using the Pearson VII function allowed us to clean the spectra obtained, and thus obtain the single scattering distribution (SSD) (blue line).

Sample	Species chemicals		Pico BE (eV)	Pico BE (eV)	Compounds	X_{CuO}	I_1/I_2	A%	SF
A-NFs	Cu^{+1}	O^{-2}	933.36	529.3	Cu^{+1} - O^{-2}	0.33–0.54	0.35–0.64	0.98	$Cu_{1.02}O_{1.15}$
	Cu^{+2}	O^{-2}	934.91	531.33	Cu^{+2} - O^{-2}	0.33–0.57	1.05–0.64	0.98	$Cu_{2.06}O_{2.03}$
Reference	Cu^{+1}	O^{-2}	933.33	529.76	Cu^{+1} - O^{-2}	0.35–0.48	0.35–0.51	0.98	$Cu_{0.97}O_{1.02}$
	Cu^{+2}	O^{-2}	936.06	530.66	Cu^{+2} - O^{-2}	0.81–0.54	1.66–1.36	0.98	$Cu_{2.43}O_{2.01}$

Note: the data obtained in SF, were obtained from calculating λ = 1.25 (Electron attenuation factor in the LEA sample), this factor was used to calculate the chemical composition by means of the MML model which was proposed as a bulk model, without layers. X_{CuO} is the Cation compositional factor, I_1/I_2 are the intensities of these, A% is the atomic percentage and SF is the final stoichiometric factor.

Table 4.
Primary XPS signal data for Cu 2p and O 1s spectra used to calculate the chemical composition in the nanopowders.

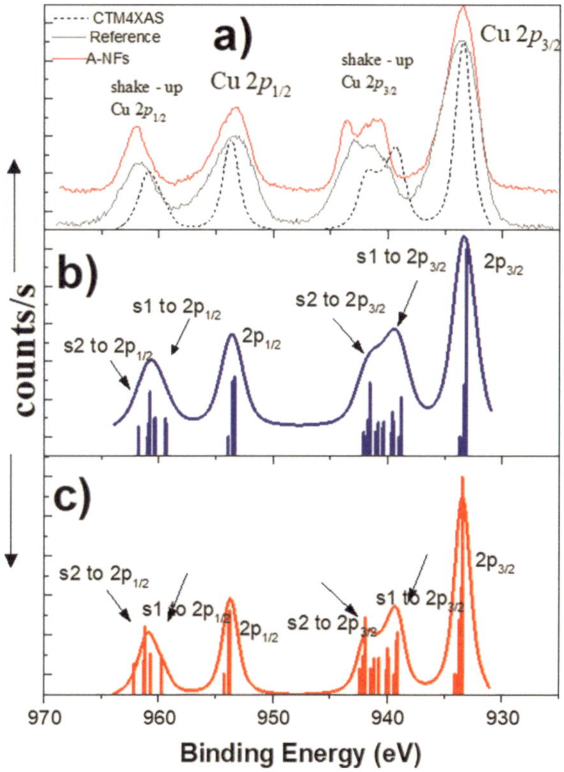

Figure 7.
Multiplet structure for the A-NFs and the reference (sigma Aldrich), compared to the calculated one (panel a). In addition, in panels b and c it differentiates the O_h and D_{4h} structures.

In addition, the relative thickness calculated by the Log-ratio method yielded values of: 35.55 (A-NFs) and 38.72 nm (B-NFs) suitable for analysis.

3.6.1 Surface and bulk plasmon analysis

Egerton reports plasmons for all materials between 10–30 eV [35], while Charles, 1996 by means of an analysis of metals and EELS analysis determined an energy of 0 to 10 eV for these metals [35]. In the study materials, plasmons were found to be associated at 22.52 and 22.15 eV, for materials A and B, respectively. Material B showed a signal of 10 eV which is consistent with being doped.

Figure 7 (Panels a,b) show the collective excitations of electrons (plasmons) where they interact in low regions in the BC and BV, known as bulk plasmon and surface plasmon, respectively. In this regard, the authors Herrera-Pérez et al., Egerton, [35, 36] propose to evaluate the energy of the free electron plasmon as:

$$E_p = 28.82\ \mathrm{eV} * \left(\frac{z\rho}{A}\right)^{1/2} \tag{5}$$

Where z is the number of electrons per molecule, ρ is the density of A-NFs = 6.520 g/ cm^3 and B-NFs = 6.473 g/cm^3 (these values were calculated by Fullprof suite software), respectively and A is the molecular weight 79.57 and 134.483 g/mol, respectively.

According to the calculations proposed by the authors Herrera-Pérez et al., Egerton [35, 36] the plasmon energy for A-NFs and C-NFs turn out to be: E_p = 16.49 eV and E_p = 17.88 eV, respectively. The change between these values is due to the interband transition below the plasmon peak [35].

These results suggest that the plasmons found are transiting in low regions between the valence and conduction bands. According to Egerton, this would indicate that they are semiconducting materials [35].

3.6.2 Comparative analysis of the dielectric function and the edges obtained from XPS-survey

Figure 8 shows the dielectric function, the peaks associated with the interband transitions obtained in EELS compared to the XPS survey spectra, for the A-NFs and C-NFs. This comparison indicates according to the author Meyer and co-workers [63], the data obtained showed a similarity.

ε_1 is the real part of the dielectric function which presents in this case a negative transition indicator showing a response below the value of the semiconductive behavior [37]. In addition, the authors Johann Toudert and Rosalía Serna [64], performed a study of the effects of collective oscillations (free charges) in Ag and Au oxides to know the optical and interband and intraband transition effects by means of the complex dielectric function ($\varepsilon = \varepsilon_1 + i\varepsilon_2$) [64]. The discussion focuses on the plasmonic effect, because on electromagnetic radiation, especially the infrared and UV-Vis zones. The results of those authors show similarity to those presented for the real function (see panels a and b), ε_1 being negative suggesting that it has the metallic character [68–71]. However, the relationship does not comply with what is exposed by these authors because $\varepsilon_1 > 1$ so such harmonic dispersion is in the UV-Vis region, but it complies with the imaginary relationship due to its growth [66]. In addition, the authors mention that this effect is for alternative plasmonic materials. In studies with doped materials, they improve the response to such an effect [72]. The author Manuel Cardona, 1986, finds that the negative value for ε_1 is related to excitons (phonon-electron) [67].

On the other hand, the peaks of the imaginary function (ε_2) were associated with the Cu 2p, 3d, and O 2p transition interbands in the conduction band, and the peaks A, D, E, H, and I are associated with the holes. In Figure 26 of reference Meyer et al. [63] present the imaginary part of the dielectric function pointing out five transition states, associated with the Cu 2*p*, 3*d*, O 2*p*, and holes bands.

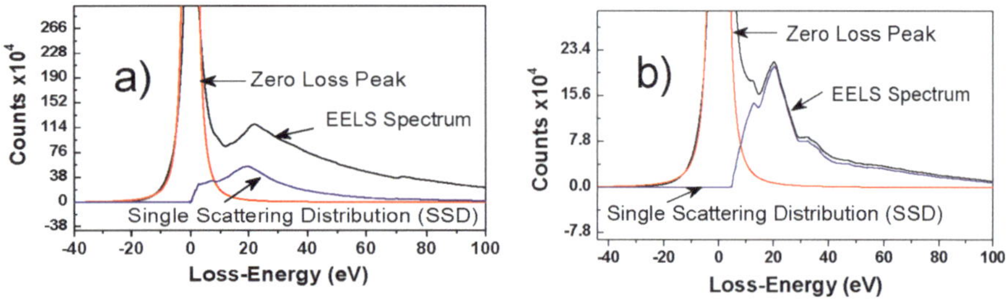

Figure 8.
Valence energy loss spectra (VEELS) of A-NFs (a) and B-NFs (b), extracted from the zero peak and dispersion spectrum (SSD).

Moreover, the bands obtained by XPS in the UPS (Ultraviolet Photoelectronic Spectroscopy) region (see panels e and f) show similarity to those reported in CuO [41, 46, 67]. Furthermore, Wang and co-workers, [73], performed the theoretical and experimental study of CuO by XPS- UPS, finding that the obtained peaks corresponded to the Cu 2p, 3d and O 2p orbitals [46, 67]. On the other hand, panel f shows the oxidative evolution of metallic Cu to the study oxides. In the study materials the leading edge presented a Cu-2p and Cu-3d splitting, while the O-2p shows an overlap with these orbitals (A-NFs and Reference), while for doping there is slightly a peak associated with it (hybridization), which agrees with the model of Wang et al. [73]. In addition, in the case of metallic Cu, the peak corresponding to oxygen was not present.

Table 5 shows the values of the electron transition shifts in the conduction band. When Mn is added to the NFs, it is observed that the main peak shows a smaller peak width (0.25 eV) and a smaller number of holes. It is also seen that the number of peaks for the A-NFs is like the theoretical one. The Cu 3d transitions for the study materials with respect to O 2p, mainly, are exposed. In **Figure 8c** and **d** a shift from 1.33 to 3.69 eV (transition Cu *2p* → Cu *3d*, see DOS in **Table 5**) in the doping with respect to pure is shown. For the case of Cu 2p with the O 2p of materials A and C there was a shift between these peaks due to doping [74–82].

Figure 9 shows the joint density of states calculated for the interband transitions. In the A-NFs (panel a), three states corresponding to those calculated by CASTEP (panel c) were observed, which coincide with those presented by the authors [41, 46, 67, 73], furthermore, the O 2p shift coincides with that of Meyer et al. [63]. In the case of the doped material, the Cu 3d transition increased the energy (see panel b) and decreased the number of holes (see **Figure 9d**). Panel c shows the holes obtained, similar to those reported by the author cited above [63]. On the other hand, the DOS model of CuO is calculated by Meyer et al. [63]. Absike, et al., [46], by calculating the TDOS and PDOS show agreement with the results (see panels c and d) [46]. Panel d shows the agreement with the one calculated by CASTEP through the GGA-PBE (Generalized Gradient Approximation - Perdew Burke Ernzerhof) functional and the one reported by Meyer and collaborators, [69].

Sample	Transition	Assignation	EELS (eV)	UPS (eV)	DOS (eV) CASTEP GGA-PBE	DOS (eV) Wang, et al. [73]	Energía de transición $\varepsilon_2(\omega)$ Meyer et al. [63]
A-NFs	B	O 2p → Cu 2p	7.75	6.60	1.94	1.98	2.71
	C	O 2p → Cu 3d	10.73	12.88	2.56	2.96	2.89
	J	O 2p	30	25.46	2.47		5.24
B-NFs	A	O 2p → Cu 2p	3.08	1.97			
	B	O 2p → Cu 3d	3.43	4.60			
	F	O 2p	30.14	21.49			

In addition, the DOS-CASTEP label shows the energy obtained from the model by this software and compares it with the authors mentioned in the Table.

Table 5.
Proposed inter-band transitions and compared to references ^{A}*[73],* ^{B}*[63].*

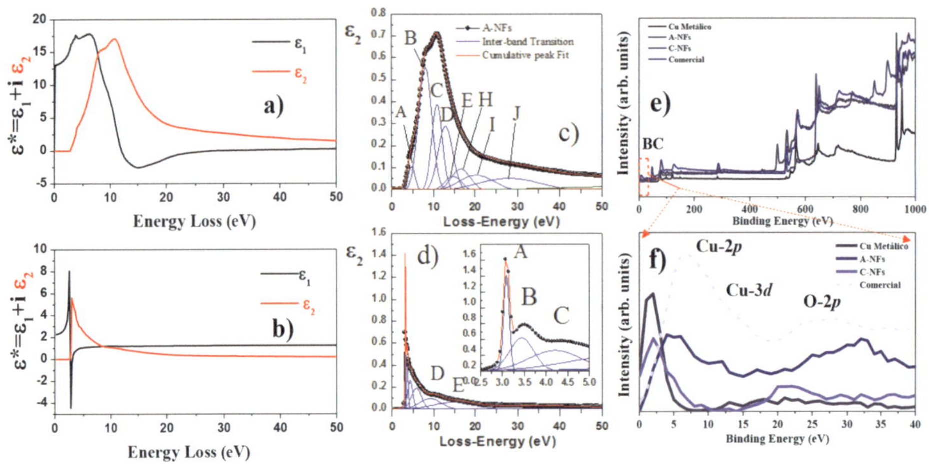

Figure 9.
Panels a and b indicate the dielectric functions $\hat{\varepsilon}(E) = \varepsilon_1(E) + i\varepsilon_2(E)$, for the A-NFs and C-NFs, respectively. In panels c and d, the interband transitions are exhibited. Panel e shows the survey spectra of: Cu metallic, A-NFs, B-NFs, and the reference. Panel f shows the Cu 3d and O 2p bands for the materials in the UPS region.

3.6.3 Bandgap determination

Figure 10 compares the normalized ELF (Energy Loss Function) (black dotted line) calculated using the KKA (Kramers-Kronig Analysis) logarithm in the 0–10 eV range with the VEELS spectra using the difference method (red line). This method allows us to calculate the onset energy associated with the (optical) bandgap in the ELF. The Eg energy was determined using the method of Rafferty and Brown [38] as the intercept of a polynomial fit $A + (E - E_g)^n$ (A is the background level), the plots suggest an indirect Eg of 2.03 and 2.85 eV, respectively. The Eg reported by Meyer et al. is 1.0 eV [69], which we attribute to working with bulk material. The calculated bandgap and that obtained in our materials suggest that the bandgap values may be associated with electronic resonance problems at the Cu 3*d* and Cu 2*p* levels due to its nanometer character, which generates changes in the orbitals, for this reason, the curve was not entirely possible to smooth it [26, 28].

3.7 Magnetic analysis

Figure 11 shows the paramagnetic behavior shown by the magnetic susceptibility of A-NFs and C-NFs (panel b). Mariammal et al. [83] performed a study with CuO and Mn-doped CuO in nanoflakes, which suggests the coupling of Mn on CuO creates spin decompensation at the surface (superparamagnetism) and spin-glass behavior that the antiferromagnetic phase disappears. According to Zhao et al., possible couplings can appear when CuO is doped with Mn, are: Cu-O-Cu-O-Mn-O-Cu-O-Cu-O-Cu, Mn-O-Mn, and Mn-O-Cu-O-Mn attributed to paramagnetic or spin-glass, antiferromagnetic, and ferromagnetic behaviors, respectively [70]. On the other hand, at d levels of Mn a strengthening in the magnetic aspect was expected, however, the antiferromagnetic effect did not occur, as in the case of NPs mentioned by Zhao et al. [70], above.

Figure 12a and **b** show the magnetization versus temperature (ZFC-FC) results of A-NFs and B-NFs pointing to temperatures of 58 and 60 K, corresponding to

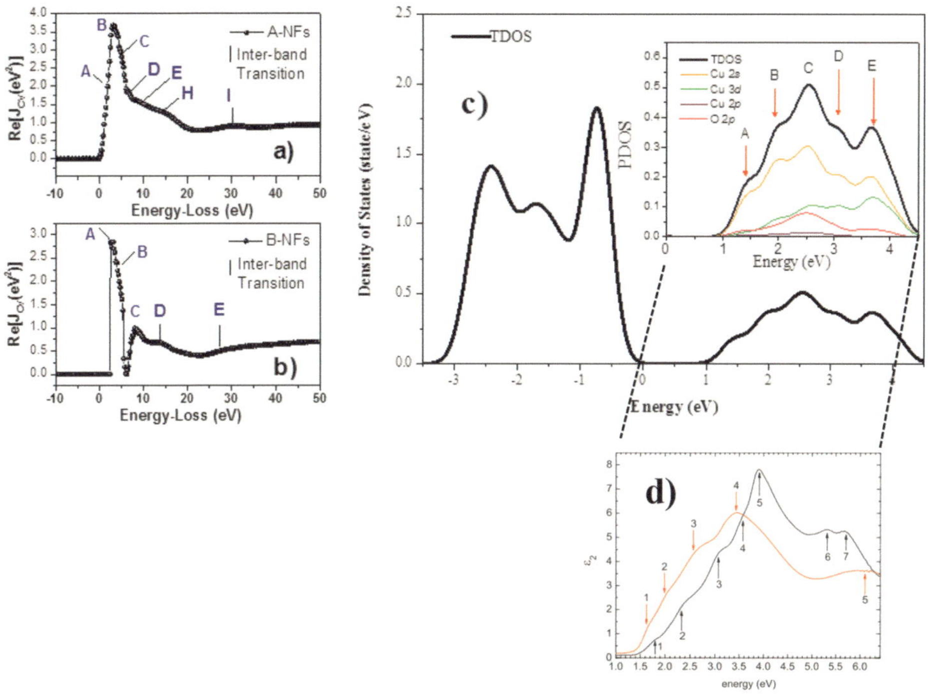

Figure 10.
Panels a and b show the joint density of states for A-NFs and B-NFs, respectively. Panel c exhibits the DOS for CuO calculated by CASTEP. In addition, the BC region with the interband transitions is exposed. Panel d shows the interband region obtained by Mayer and the gaps obtained from this model [69].

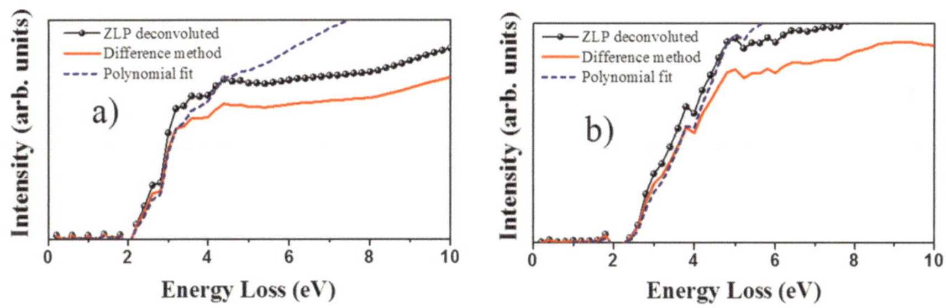

Figure 11.
In panels a-b, the comparisons of the VEELS spectra with deconvoluted ZLP and the fit of these with the difference method are shown. In addition, the blue dashed line indicates the polynomial fit for E_g. a) Corresponds to the A-NFs and b) are the B-NFs.

ferromagnetic behavior. This behavior is also observed in the results of NPs in CuO and Mn-doped CuO (nanowires) of the author Han et al. [12] shown in panel c. Similar to this work, it shows a blocking T greater than 80 K which is attributed to the different grain size and higher doping than this work.

Recalling that bulk CuO exhibits two antiferromagnetic phases at 213 and 230 K it is remarkable that the same in **Figure 12a** and **b** no such phases are observed in both cases due to spin decompensation at the surface of the nanostructures [84]. The absence of these phases also in the work of Narsinga Rao et al., [71] in nanoparticles shown in the image d.

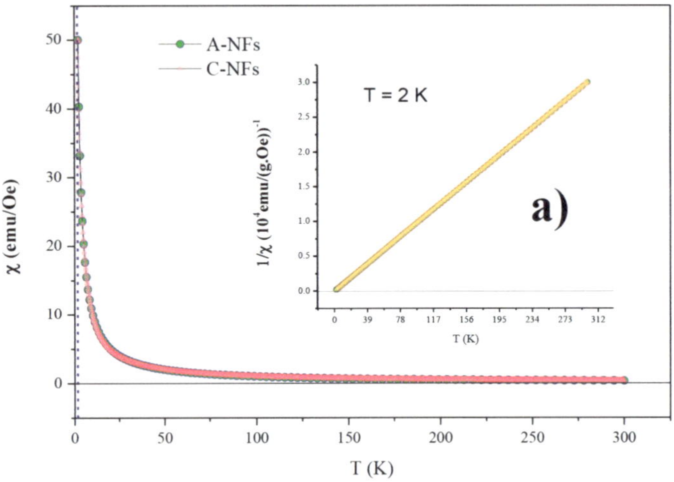

Figure 12.
Magnetic susceptibility from 2 K to 300 K, a) for CuO NFs, pure and doped with Mn (2.5%).

Figure 13a and **b** show the magnetic behaviors using hysteresis loops for A-NFs and B-NFs at 300 K, which were found to be superparamagnetic and paramagnetic, respectively. The superparamagnetic behavior is comparable with that obtained for NPs by Narsinga Rao et al. [71].

In **Table 6**, the data of magnetic saturation variations and coercivity for the synthesized materials and CuO NPs of authors Borzi et al. [85] and Narsinga Rao et al.

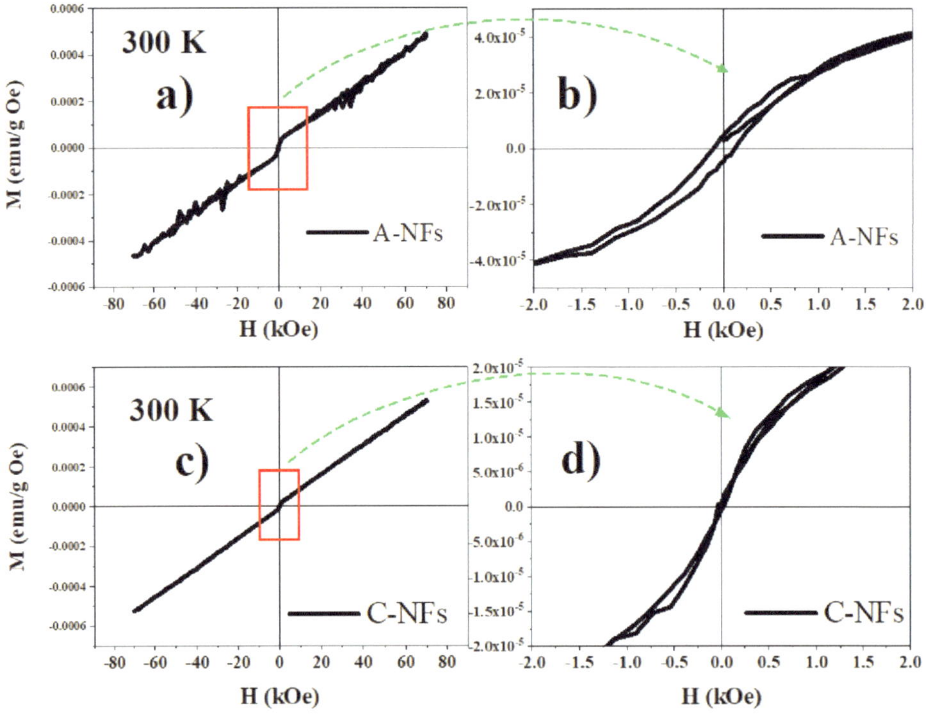

Figure 13.
Magnetic analysis panel (a and (b for A-NFs and panel c) and d) for B-NFs at 300 K.

Sample	H_c (KOe)	M_r (emu/g)
A-NFs	0.25	2.80×10^{-5}
B-NFs	0.006	0.002
Borzi, et al. [85]	0.200	0.007
Narsinga Rao et al. [71]	0.200	0.306

Table 6.
Coercivity and Remanence for A and B materials, compared with Borzi et al. and Narsinga Rao et al. [71, 85].

[71] at room T are shown. It is observed that the coercivity is higher in A-NFs than those reported by those authors. The difference is attributed to the nanometer dimensions between the nanostructures handled and the difference in the geometries between the NFs and the NPs of these mentioned authors. On the other hand, C-NFs present higher remanence with respect to A-NFs, but also lower coercivity.

4. Conclusions

The synthesis of A-NFs by electrospinning showed a monoclinic phase corresponding to tenorite. These NFs show multivalent states (Cu^{1+}, Cu^{2+} and Cu^{3+}). Their lattice parameters calculated by the Rietveld method suggest lattice distortion. This was corroborated by the electron density in the crystal structure with charge transfer from O^{2-} to Cu^{2+} appreciated as peanut shape which is characteristic in semiconducting materials [1].

The XPS study showed the identification of the cations/anions in the Cu 2p and O 1s orbitals, corroborated with the authors [1–7], respectively. The distances of the main branches (Cu $2p_{3/2}$ and Cu $2p_{1/2}$) were: 19.97 and 19.91 eV, respectively, because of the Jahn-Teller effect [6, 86, 87]. The increase in the distance of the main branches is a consequence of paramagnetism in CuO. The spectra were used to calculate the stoichiometry using the geometrical topofactor (sphere) method showing non-stoichiometric oxides: $Cu_{1.02}O_{1.15}$ and $Cu_{2.06}O_{2.03}$, and the reference: $Cu_{0.97}O_{1.02}$ and $Cu_{2.43}O_{2.01}$. The CTM4XAS software corroborated the cationic states in the Cu $2p$ orbital through the multiplet structure. In addition, the calculation employed a higher charge transfer than previous reports. Moreover, the multiplet structure of A-NFs suggests a D_{4h} structure presents higher number of holes (this chemical coordination is suggested as the one with the lowest energy according to the experimental spectral shape). On the other hand, the O $1s$ spectra allowed us to calculate the oxygen vacancies generated in these, found in the same samples at 533.17, 532.22 eV. These energies agree with previous reports.

The difference method in VEELS spectra allowed to calculate the onset energy associated with Eg (optical) using the energy loss function (EFL). In addition, this method suggests the elimination of the relativistic effect of the electrons (10^6) from the spectra [13, 14]. The polynomial fit for the spectra presented an indirect Eg, in agreement with that calculated by CASTEP agrees and differs with previous reports.

The comparative XPS-EELS study for A-NFs and B-NFs (3.0% Mn-doped CuO) in the UV-VIS region (0–50 eV) through the dielectric function showed similarity in spectral shape, deconvolution of the EELS spectra determined the interband states in the CB for Cu 3d, Cu 2s and O 2p orbitals. Furthermore, the comparative study of the A-NFs and B-NFs suggests changes in the interband transitions in agreement with those calculated, the E_g by CASTEP and VEELS obtained were: 1.27, 2.03, 2.85 eV. Meyer et al. [63, 69] shows.

According to the magnetization study, it was observed that sample A and B Ferromagnetism and Paramagnetic (mixed phase) behavior, which is attributed to the change obtained in the Cu 2p multiplet peaks of XPS (change in electric current arises due to changes in $\vec{J} = \vec{L} + \vec{S}$). These results are compared to those of reference [16], finding a mixed phase. This is due to the spin decompensation at the $3d^9$ orbital, which is attributed to its nanometer size and doping and B a paramagnetic behavior, which is attributed to its nanometer size and doping. In addition, the antiferromagnetic phase that normally appears in a CuO (II) bulk material at 232 K, in these samples is not present. This is attributable to the nanometer size, particle geometry, oxygen vacancies, charge decompensation (Cu^{3+}) and occurs in other nanostructures [16, 70, 88].

Acknowledgements

The author M. Piñón-Espitia thanks the Ph.D. CONACYT scholarship grant No. 467043. We would like to make two important mentions for the completion of this work: to Dr. Francesca Péiro and Dr. Luis Gerardo Silva for the support received in the EELS and XPS studies. As well as to the national laboratory of nanotechnology CIMAV, S.C. Chihuahua.

Conflict of interest

The authors have no conflicts to disclose.

Abbreviations

NFs	nanofibers
A-NFs	CuO Nanofibers
B-NFs	CuO doped with Mn at 3.0% nanofibers
NPs	Nanoparticles
SEM	scanning electron microscopy
HR-TEM	high-resolution transmission electron microscopy
XRD	X-ray diffraction
TGA-DSC	thermogravimetric analysis-Differential scanning calorimetry
XPS	X-ray photoelectronic spectroscopy
UPS	ultraviolet photoelectronic spectroscopy
CTM4XAS	charge transfer multiplet
UV–Vis	ultraviolet visible
EELS	electron energy loss spectroscopy
VEELS	valence electron energy loss spectroscopy
GGA-PBE	generalized gradient approximation- Perdew Burke Ernzerhof
CASTEP	Cambridge serial total energy package
DFT	density functional theory
DOS	density of state
AFM	antiferromagnetic
FM	ferromagnetic

Author details

Manuel F. Piñón-Espitia[1]*, Guillermo M. Herrera-Pérez[2] and Matha T. Ochoa-Lara[1]

1 Departament of Physics of Materials, Centro de Investigación en Materiales Avanzados, Chihuahua, Mexico

2 Catedra CONACYT asignada al Departamento de Físcia de Materiales, Centro de Investigación en Materiales Avanzados, Chihuahua, Mexico

*Address all correspondence to: freunfide@gmail.com; manuel.pinon@cimav.edu.mx

References

[1] Wu H, Lin D, Pan W. Electrospinning of ceramic nanofibers: Fabrication, assembly and applications. Journal of Advanced Ceramics. 2012;**1**(1):2-23. DOI: 10.1007/s40145-012-0002-4

[2] Sahai A, Goswami N, Kaushik SD, Tripathi S. Cu/Cu2O/CuO nanoparticles: Novel synthesis by exploding wire technique and extensive characterization. Applied Surface Science. 2016;**390**:974-983. DOI: 10.1016/J.APSUSC.2016.09.005

[3] Åsbrink S, Norrby LJ. A refinement of the crystal structure of copper(II) oxide with a discussion of some exceptional e.s.d.'s. Acta Crystallographica Section B. 1970;**26**(1):8-15. DOI: 10.1107/s0567740870001838

[4] Basith NM, Vijaya JJ, Kennedy LJ, Bououdina M. Structural, morphological, optical, and magnetic properties of Ni-doped CuO nanostructures prepared by a rapid microwave combustion method. Materials Science in Semiconductor Processing. 2014;**17**:110-118. DOI: 10.1016/J.MSSP.2013.09.013

[5] Torres-Ochoa JA, Mulato-Gomez DF, German-Cabrera D, Bravo-Sanchez M, Herrera-Gomez A. Detailed Analysis of the Photoemission Spectra of Copper Films with Coexistent Cu1+, Cu2+ and Cu3+ Oxidation States. 2017. pp. 125-126

[6] Sarkar S, Kumar Jana P, Chaudhuri BK, Sakata H. Copper (II) oxide as a giant dielectric material synthesis of Cu 2 O from CuO thin films: Optical and electrical properties copper " II ... oxide as a giant dielectric material. Applied Physics Letters. 2006;**8911**(10):212905-212703. DOI: 10.1063/1.2393001

[7] Sonia S, Annsi IJ, Suresh Kumar P, Mangalaraj D, Viswanathan C, Ponpandian N. Hydrothermal synthesis of novel Zn doped CuO nanoflowers as an efficient photodegradation material for textile dyes. Materials Letters. 2015;**144**:127-130. DOI: 10.1016/j.matlet.2015.01.026

[8] Yu T, Zhao X, Shen ZX, Wu YH, Su WH. Investigation of individual CuO nanorods by polarized micro-Raman scattering. Journal of Crystal Growth. 2004;**268**(3–4 SPEC. ISS):590-595. DOI: 10.1016/j.jcrysgro.2004.04.097

[9] Manjunath A, Irfan M, Anushree KP, Mahesha Vinutha K, Yamunarani N. Synthesis and characterization of CuO nanoparticles and CuO doped PVA nanocomposites. Advances in Materials Physics and Chemistry. 2016;**6**:263-273. DOI: 10.4236/ampc.2016.610026

[10] Ray A et al. Optimization of photoelectrochemical performance in chemical bath deposited nanostructured CuO. Journal of Alloys and Compounds. 2017;**695**:3655-3665. DOI: 10.1016/j.jallcom.2016.11.374

[11] Gaurav A, Beura R, Kumar JS, Thangadurai P. Study on the effect of copper ion doping in zinc oxide nanomaterials for photocatalytic applications. Materials Chemistry and Physics. 2019;**230**(March 2018):162-171. DOI: 10.1016/j.matchemphys.2019.03.056

[12] Han D, Wu Z, Wang Z, Yang S. Oriented Mn-doped CuO nanowire arrays. Nanotechnology. 2016;**27**(13):135603

[13] Xiao H-M, Zhu L-P, Liu X-M, Fu S-Y. Anomalous ferromagnetic behavior of CuO nanorods synthesized via hydrothermal method. Solid State Communications. 2007;**141**(8):431-435. DOI: 10.1016/j.ssc.2006.12.005

[14] Zheng XG et al. Finite-size effect on Néel temperature in antiferromagnetic nanoparticles. Physical Review B. 2005; **72**:014464. DOI: 10.1103/PhysRevB.72.014464

[15] Punnoose A, Magnone H, Seehra MS. Bulk to nanoscale magnetism and exchange bias in CuO nanoparticles. Physical Review B. 2001;**64**:1-8. DOI: 10.1103/PhysRevB.64.174420

[16] Gao D, Yang G, Li J, Zhang J, Zhang J, Xue D. Room-temperature ferromagnetism of flowerlike CuO nanostructures. Physical Chemistry C. 2010;**114**(43):18347-18351. DOI: 10.1021/jp106015t

[17] He H. Metal Oxide Semiconductors and Conductors. Elsevier Inc.; 2020. DOI: 10.1016/b978-0-12-814930-0.00002-5

[18] Jeong YK, Choi GM. Nonstoichiometry and electrical conduction of CuO. Journal of Physics and Chemistry of Solids. 1996;**57**(1):81-84. DOI: 10.1016/0022-3697(95)00130-1

[19] Siraja KHF, Rafiquea MS, Naseemb S. Microstructural and optical properties of zinc doped copper oxide thin films. Materials Today Proceedings. 2015;**2**: 5426-5429

[20] Cruz-Filho JF et al. Effect of different synthesis methods on the morphology, optical behavior, and superior photocatalytic performances of Ag_3PO_4 sub-microcrystals using white-light-emitting diodes. Journal of Photochemistry and Photobiology A: Chemistry. 2019;**377**(October 2018): 14-25. DOI: 10.1016/j.jphotochem.2019.03.031

[21] Sahay R, Sundaramurthy J, Suresh Kumar P, Thavasi V, Mhaisalkar SG, Ramakrishna S. Synthesis and characterization of CuO nanofibers, and investigation for its suitability as blocking layer in ZnO NPs based dye sensitized solar cell and as photocatalyst in organic dye degradation. Journal of Solid State Chemistry. 2012;**186**:261-267. DOI: 10.1016/j.jssc.2011.12.013

[22] Vinothkumar P, Manoharan C, Shanmugapriya B, Bououdina M. Effect of reaction time on structural, morphological, optical and photocatalytic properties of copper oxide (CuO) nanostructures. Journal of Materials Science: Materials in Electronics. 2019;**30**(6):6249-6262. DOI: 10.1007/s10854-019-00928-7

[23] Pramothkumar A, Senthilkumar N, Malar KCMG, Meena M, Potheher IV. A comparative analysis on the dye degradation efficiency of pure, Co, Ni and Mn-doped CuO nanoparticles. Journal of Materials Science: Materials in Electronics. 2019;**30**(20):19043-19059. DOI: 10.1007/s10854-019-02262-4

[24] Aiswarya Raj VBAS. Nanostructured CuO: Facile synthesis, optical absorption and defect dependent electrical conductivity. Materials Science in Semiconductor Processing. 2017; **68**:38-47

[25] Okada K, Kotani A. Multiplet structure in core - level XPS of high - Tc material La_2CuO_4 and related Cu - compounds. Journal of Electron Spectroscopy and Related Phenomena. 1990;**52**:313-322

[26] Piñon-Espitia M, Lardizabal-Gutiérrez D, Camacho-Ríos ML, Herrera-Pérez G. Electronic structure comparison of Cu 2p and O 1s X-Ray photoelectron spectra for Cu x O nanofibers (x = 1, 2, i). Materials Chemistry and Physics. 2021;**272** (December 2020):124980-124987. DOI: 10.1016/j.matchemphys.2021.124981

[27] Herrera-Gomez A, Aguirre-Tostado F-S, Pianetta P. Formation of Si1+ in the early stages of the oxidation of the Si [001] 2 x 1 surfaces. Journal of Vacuum Science & Technology. 2016;**34** (020601):1-4

[28] Herrera-Gomez A et al. Intensity modulation of the Shirley background of the Cr 3p spectra with photon energies around the Cr 2p edge. Surface and Interface Analysis. 2018;**50**(2):246-252. DOI: 10.1002/sia.6364

[29] Cabrera-German D, Gomez-Sosa G, Herrera-Gomez A. Accurate peak fitting and subsequent quantitative composition analysis of the spectrum of Co 2p obtained with Al Kα radiation: I: Cobalt spinel. Surface and Interface Analysis. 2016;**48**(5):252-256. DOI: 10.1002/sia.5933

[30] Salvia AM, Castle JE. The intrinsic asymmetry of photoelectron peaks: Dependence on chemical state and role in curve fitting. Journal of Electron Spectroscopy and Related Phenomena. 1998;**95**:45-56

[31] NIST X-ray Photoelectron Spectroscopy (XPS) Database Main Search Menu. Available from: https://srdata.nist.gov/xps/main_search_menu.aspx [Accessed: October 10, 2017]

[32] Stavitski E, De Groot FMF. Author's personal copy the CTM4XAS program for EELS and XAS spectral shape analysis of transition metal L edges. Micron. 2010;**41**:687-694

[33] Piñon-Espitia M, Lardizabal-Gutiérrez D, Camacho-Ríos ML, Herrera-Pérez G, Duarte-Moller A, Ochoa-Lara MT. Charge transfer effects and O_2-vacancies in pure CuO nanofibers and enriched with 3.0% Mn. Materials Chemistry and Physics. 2022; **295**(August 2021):126989. DOI: 10.1016/j.matchemphys.2022.126989

[34] Park J, Yang M. Determination of complex dielectric functions at HfO_2/Si interface by using STEM-VEELS. Micron. 2009;**40**(3):365-369. DOI: 10.1016/j.micron.2008.10.006

[35] Egerton RF. Electron Energy-Loss Spectroscopy in the Electron Microscope. Springer Science & Business Media; 2009

[36] Herrera-Pérez LF-CG, Ornelas-Gutiérrez C, Reyes-Montero A, Paraguay-Delgado F, Reyes-Rojas A. Complex dielectric function and opto-electronic characterization using VEELS for the lead-free BCZT electro-ceramic perovskite. Micron. 2021;**149**(March): 103124-103130. DOI: 10.1016/j.micron.2021.103124

[37] Eljarrat A. Optoelectronic properties of InAlN/GaN distributed Bragg reflector heterostructure examined by valence electron energy loss spectroscopy microscopy microanalysis optoelectronic properties of InAlN/GaN distributed Bragg reflector heterostructure exam. Microscopy and Microanalysis. 2012;**18**(5):1143-1154. DOI: 10.1017/S1431927612001328

[38] Rafferty B, Brown LM. Direct and indirect transitions in the region of the band gap using electron-energy-loss spectroscopy. Physical Review B. 1998; **58**(16):326-337

[39] Rodriguez-Carvajal J. Fullprof. Physics B. 1993;**192**:55

[40] Momma FIK. VESTA 3 para visualización tridimensional de cristal, datos volumétricos y mofológicos. Journal of Applied Crystallography. 2011;**44**:1272-1276

[41] Shard AG, Wang J, Spencer SJ. XPS topofactors: Determining overlayer

thickness on particles and fibres. Surface and Interface Analysis. 2009;**41**:541-548

[42] Bravo Sanchez M, Huerta-Ruelas JA, Cabrera-German D, Herrera-Gomez A. Composition assessment of ferric oxide by accurate peak fitting of the Fe 2p photoemission spectrum. Surface and Interface Analysis. 2017;**49**(4):253-260. DOI: 10.1002/sia.6124

[43] Ghijsen J et al. Electronic structure of Cuzo and Cuo. Physical Review B. 1988;**38**(16):11322

[44] Origin(Pro), Version 2011. OriginLab Corporation, Northampton, MA, USA

[45] Ikeno H, de Groot FMF, Stavitski E, Tanaka I. Multiplet calculations of L 2,3 x-ray absorption near-edge structures for 3d transition-metal compounds. Journal of Physics. Condensed Matter. 2009;**21**:17. DOI: 10.1088/0953-8984/21/10/104208

[46] Absike H, Hajji M, Labrim H, Abbassi A, Ez-Zahraouy H. Electronic, electrical and optical properties of Ag doped CuO through modified Becke-Johnson exchange potential. Superlattices and Microstructures. 2019;**127**:128-138. DOI: 10.1016/j.spmi.2017.12.038

[47] Kröger FA, Vink HJ. Relations between the concentrations of imperfections in crystalline solids. Solid State Physics. 1956;**3**(C):307-435. DOI: 10.1016/S0081-1947(08)60135-6

[48] Piñón-Espitia M, Torres-Ochoa J, Córtazar-Martínez O, Herrera-Gomez A, Herrera-Perez G, Ochoa-Lara MT. Charge transfer effects in CuO nanofibers, observed by EELS and XPS. Microscopy and Microanalysis. 2020;**26**(S2):1424-1425. DOI: 10.1017/s143192762001805x

[49] Cao H, Zhou Z, Yu J, Zhou X. DFT study on structural, electronic, and optical properties of cubic and monoclinic CuO. Journal of Computational Electronics. 2018;**17**(1):21-28. DOI: 10.1007/s10825-017-1057-9

[50] Sarkar S, Jana PK, Chaudhuri BK, Sakata H. Copper (II) oxide as a giant dielectric material. Applied Physics Letters. 2006;**89**(21):1-4. DOI: 10.1063/1.2393001

[51] Bahgat AA, Shaisha EE, Saber MM. Study of microstructure and magnetic properties in copper oxide superconducting systems through AC magnetic susceptibility. Physica B: Condensed Matter. 2007;**399**(1):70-76. DOI: 10.1016/j.physb.2007.05.027

[52] Steiner P et al. The Cu valence in the high Tc superconductors and in monovalent, divalent and trivalent copper oxides determined from XPS core level spectroscopy. Zeitschrift für Physik B Condensed Matter. 1987;**67**(4):497-502. DOI: 10.1007/BF01304119

[53] Pauly N, Tougaard S, Yubero F. Determination of the Cu 2p primary excitation spectra for Cu, Cu_2O and CuO. Surface Science. 2014;**620**:17-22

[54] Duque Sánchez LM, Rodriguez L, López M. Electrospinning: La era de las nanofibras. Revista Iberoamericana de Polímeros. 2014;**14**(1):10-27

[55] Cabrera N, Mott NF. Theory of the oxidation of metals. Reports on Progress in Physics. 1949;**12**(1):308. DOI: 10.1088/0034-4885/12/1/308

[56] Chusuei CC, Brookshier MA, Goodman DW. Correlation of relative X-ray photoelectron spectroscopy shake-up intensity with CuO particle size.

Langmuir. Apr 1999;**15**(8):2806-2808. DOI: 10.1021/la9815446

[57] Chang CL, Lee TC, Huang TJ. High performance copper oxide cathodes for high temperature solid-oxide fuel cells. Journal of Applied Electrochemistry. 1996;**26**(3):311-317. DOI: 10.1007/BF00242101

[58] Zeng J, Xu J, Wang S, Tao P, Hua W. Ferromagnetic behavior of copper oxide-nanowire-covered carbon fibre synthesized by thermal oxidation. Materials Characterization. 2009;**60**: 1068-1070. DOI: 10.1016/j.matchar.2009.03.012

[59] Bagus PS, Ilton ES, Nelin CJ. The interpretation of XPS spectra: Insights into materials properties. Surface Science Reports. 2013;**68**:273-304. DOI: 10.1016/j.surfrep.2013.03.001

[60] Bagus PS, Nelin CJ, Al-Salik Y, Ilton ES, Idriss H. Multiplet splitting for the XPS of heavy elements: Dependence on oxidation state. Surface Science. 2016; **643**:142-149. DOI: 10.1016/j.susc.2015.06.002

[61] Shard AG. A straightforward method for interpreting XPS data from core–shell nanoparticles. Journal of Physical Chemistry C. 2012;**116**(31):16806-16813. DOI: 10.1021/jp305267d

[62] Qin H, Zhang Z, Liu X, Zhang Y, Hu J. Room-temperature ferromagnetism in CuO sol-gel powders and films. Journal of Magnetism and Magnetic Materials. 2010;**322**(14):1994-1998. DOI: 10.1016/j.jmmm.2010.01.021

[63] Meyer BK et al. Binary copper oxide semiconductors: From materials towards devices. Physica Status Solidi. Aug 2012; **249**(8):1487-1509. DOI: 10.1002/pssb.201248128

[64] Toudert J, Serna R. Interband transitions in semi-metals, semiconductors, and topological insulators: A new driving force for plasmonics and nanophotonics [invited]. Optical Materials Express. 2017;7(7): 2299-2325

[65] Maier SA. Plasmonics: Fundamentals and Applications. Vol. 30. Springer; 2007. pp. 65-88. DOI: 10.1007/0-387-37825-1

[66] Fischer DBMP, Schmidt C, Sakat E, Stock J, Samarelli A, Frigerio J, et al. Optical activation of germanium plasmonic antennas in the mid-ifrared. Physical Review Letters. 2016;**117**(4): 047401

[67] Cardona M. Temperature dependence of the dielectric function and the interband critical points of CdSe. Physical Review B. 1986;**34**(4):2458

[68] Tahir D, Tougaard S. Electronic and optical properties of Cu, CuO and Cu_2O studied by electron spectroscopy. Journal of Physics. Condensed Matter. 2012;**24**: 175002. DOI: 10.1088/0953-8984/24/17/175002

[69] Meyer CRBK, Polity A, Reppin D, Becker M, Hering P, Klar PJ, et al. Binary copper oxide semiconductors: From materials towards devices. Basic Solid State Physics. 2012;**249**:1487-1509

[70] Zhao F et al. Ferromagnetism analysis of Mn-doped CuO thin films. Journal of Physics. Condensed Matter. 2008;**20**:425208-425205. DOI: 10.1088/0953-8984/20/42/425208

[71] Narsinga Rao G, Yao YD, Chen JW. Evolution of size, morphology, and magnetic properties of CuO nanoparticles by thermal annealing. Journal of Applied Physics. 2009;**105**(9): 093901. DOI: 10.1063/1.3120785

[72] Luther APAJM, Jain PK, Ewers T. Localized surface plasmon resonances arising from free carriers in doped quantum dots. Nature Materials. 2011;**10** (5):361-366

[73] Wang Y et al. Electronic structures of Cu_2O, Cu_4O_3, and CuO: A joint experimental and theoretical study. Physical Review B. 2016;**94**(24):245418. DOI: 10.1103/PhysRevB.94.245418

[74] Sharma N, Gaur A, Kotnala RK. Signature of weak ferroelectricity and ferromagnetism in Mn doped CuO nanostructures. Journal of Magnetism and Magnetic Materials. Mar 2015;**377**: 183-189. DOI: 10.1016/J.JMMM. 2014.10.055

[75] Wang W, Zhang L, Tong S, Li X, Song W. Three-dimensional network films of electrospun copper oxide nanofibers for glucose determination. Biosensors & Bioelectronics. 2009;**25**: 708-714. DOI: 10.1016/j. bios.2009.08.013

[76] Cho HJ, Onozato T, Wei M, Sanchela A, Ohta H. Effects of vacuum annealing on the electron mobility of epitaxial La-doped $BaSnO_3$ films. APL Materials. 2019;7(2). DOI: 10.1063/1.5054154

[77] Ghijsen J, Tjeng LH, Eskes H, Sawatzky GA, Johnson RL. Resonant photoemission study of the electronic structure of Cuo and Cu_20. Physical Review B: Condensed Matter. 1990;**42** (4):2268-2274

[78] van der Laan AGASG, Westra C, Haans C. Satellite structure in photoelectron and auger spectra of copper dihalides. Physical Review B. 1981;**23**:4369

[79] Jorgensen JD, Radaelli PG, Shaked H, Wagner JL, Hunter BA, Mitchell JF, et al. Roles of oxygen defects in copper oxide superconductors. Journal of Superconductivity. 1994;**7**(1):1-5

[80] Yaqub M et al. Role of multivalent Cu, oxygen vacancies and CuO nanophase in the ferromagnetic properties of ZnO: Cu thin film. RSC Advances. 2015;**20**:1-10. DOI: 10.1039/ C6RA12827K

[81] Kreibig U, Vollmer M. Optical Properties of Metal Clusters. Vol. 25. Springer; 1995. DOI: 10.1007/978-3-662-09109-8

[82] Ross MB, Schatz GC. Aluminium and indium plasmonic nanoantennas in the ultraviolet. Physical Chemistry C. 2014;**118**(23):12506-12514

[83] Mariammal RN, Ramachandran K, Kalaiselvan G, Arumugam S, Renganathan B, Sastikumar D. Effet of magnetism on the ethanol sensitivity of undoped and Mn - doped CuO nanoflakes. Applied Surface Science. 2013;**270**:545-552. DOI: 10.1016/j.apsusc.2013.01.084

[84] Otero J. Estudio del superparamagnetismo: nanomagnetismo en partículas Trabajo de la asignatura. In: Introducción a la Ciencia de Autor : Javier Otero Márquez. Scotland: University of Edinburgh; 2015. DOI: 10.13140/RG.2.1.1738.7362

[85] Borzi RA, Stewart SJ, Mercader RC, Punte G, Garcia F. Magnetic behavior of nanosized cupric oxide. Journal of Magnetism and Magnetic Materials. 2001;**226–230**(PART II):1513-1515. DOI: 10.1016/S0304-8853(00)00943-4

[86] Cullity BD, Graham CD. Introduction to Magnetic Materials. IEEE/Wiley; 2009

[87] Bhardwaj R, Bharti A, Singh JP, Chae KH, Goyal N. Influence of Cu doping on

the local electronic and magnetic properties of ZnO nanostructures. Nanoscale Advances. 2020;**2**:4450-4463. DOI: 10.1039/d0na00499e

[88] Rao GN, Yao YD, Chen JW. Superparamagnetic behavior of antiferromagnetic CuO nanoparticles. IEEE Transactions on Magnetics. 2005; **41**(10):3409-3411. DOI: 10.1109/TMAG.2005.855214

Chapter 5

Development of Drug-Delivery Textiles Using Different Electrospinning Techniques: A Review

Isabel C. Gouveia and Cláudia Mouro

Abstract

Electrospinning, a remarkable and versatile technique has been related to medical textiles, aiming to produce nanomaterials for drug delivery and tissue regeneration applications. Furthermore, electrospun nanofibrous materials with unique properties as favorable pore size distribution, porosity, surface area, and wettability, along with effective mechanical properties, are the frontrunner solutions. Also, the features of the nanofibrous structures can be designed and optimized by controlling electrospinning parameters related to the solution properties, the setup parameters, and the environmental conditions to design nanofibrous textile materials for the desired applications. Further, to accomplish the required functionality of the drug-delivery systems, a rather broad range of drugs have been loaded into the nanofibers using different electrospinning techniques, namely the blending, side-by-side, coaxial, tri-axial, emulsion, and multi-needle electrospinning, in order to accomplish specific drug-release profiles of the designed nanofibrous textiles. Thus, this chapter describes the different electrospinning techniques that have been utilized in the production of the textile nanofibrous materials as the application of these materials in bone, nerve, periodontal, and vascular regeneration, as well as in wound dressings, personal-protective-equipment (PPE), and cancer treatment, providing an overview of the recent studies and highlighting the current challenges and future perspectives for their medical applications.

Keywords: emulsion electrospinning (ES), coaxial ES, blend ES, encapsulation ES, drug delivery, medical textiles, wound dressing, personal protective equipment

1. Introduction

During the last years, the huge and rapid development in nanotechnology has greatly contributed to the production of new functional materials with enhanced properties to be effectively and efficiently used in numerous fields [1–3]. Nevertheless, the development of medical textiles through the intervention of nanotechnology has been highlighted, in order to improve the health and quality of life of patients [3, 4].

Medical textiles, due to their distinctive features, such as their flexibility, wide range of sizes, and light-weight, physical, structural, and surface properties, have been produced with different materials (synthetic and/or natural) and with various structures (woven, knitted, braided, nonwoven, and composites) for different purposes, namely for advanced biomedical applications, like drug-delivery systems for tissue regeneration and wound healing [4–6]. However, developing advanced biocompatible, biodegradable, nontoxic, and nonallergic materials with desirable properties for drug loading and delivery in biomedical fields is still one the major challenges faced by the researchers, and thus, to minimize this gap, textile-based drug-delivery materials have been designed [3, 6, 7].

Among the different types of medical textiles, electrospun textiles fabricated from natural polymers, like proteins and polysaccharides, have gained increasing interest due to their low cost, biocompatibility, biodegradability, bioactivity, and unique properties, like the high surface area-to-volume ratio, smaller and more uniform fiber diameters, highly interconnected porous structures, and similarity to the extracellular matrices (ECMs) of tissues in the body [3, 8–11]. Moreover, electrospinning has been revealed to be a simple and versatile method to produce drug-delivery systems, and hence the nanofibrous materials developed from the electrospinning are a unique and exceptional platform for loading and delivery of multiple therapeutic agents due to their performance and intrinsic nanoscale morphological characteristics [3, 8–11]. Furthermore, the nanofibers drug loading has been achieved using different electrospinning techniques that comprise the incorporation of these therapeutic agents prior to the electrospinning process through the blended, coaxial, side-by-side, emulsion, and tri-axial electrospinning in order to deliver the appropriate amount of drug over the desired period of time and with a specific release profile [3, 5, 11]. Therefore, textile-based drug loading and delivery materials have been designed according to the medical conditions in order to improve the effectiveness of the drugs and reduce costs and the toxic side effects [5].

In the following sections of this chapter, an overview is provided which focuses on the fundamentals of the electrospinning technology, the different parameters that influence the diameter and arrangement of the produced nanofibers, and the main medical textile materials (e.g., cellulose, chitosan, collagen, alginate, keratin, and silk) used to produce textile-based drug loading and delivery nanofibers. In addition, the drug-release mechanisms and kinetics, as well as the different electrospinning techniques used to incorporate different therapeutic agents and produce electrospun nanofibers to act as drug-delivery textile materials, are described in detail. Finally, a brief outline of the recent studies concerning the production of textile-based drug loading and delivery materials for medical fields, namely for applications in bone, nerve, periodontal, and vascular grafts tissue engineering, wound dressings, and other textile-based materials, like personal protective equipment (PPE) and cancer treatment is presented, as well as the conclusions, challenges, and future perspectives of this emerging research field.

2. Electrospinning technology to produce textile-based drug-delivery materials

Electrospinning is a simple, versatile, and cost-effective manufacturing method that uses high voltage to produce an electrically charged jet of polymer toward the collector, where the continuous nanofibers are collected and interconnected nanofibrous structures produced (**Figure 1**) [3, 5, 8, 9, 12–15]. Therefore, in the electrospinning process, a charged polymer jet is ejected when the electrostatic forces overcome the surface tension of the polymer solution and a Taylor cone-jet, subjected to a variety of forces, like an electric force imposed by the external electric field, a Coulomb force, a surface tension force, a viscoelastic force, and gravitational forces, is formed to produce the high-quality nanofibers [3, 8, 9, 12–14].

In addition, the electrospun textiles have been recognized by their superior structural properties, like the high surface area-to-volume ratio, tunable porosities, morphologies, and diameters, as well as the ability to carry large amounts of drugs or to be chemically modified using wet chemical methods, oxygen plasma treatments, and in situ grafting polymerization for specific applications, such as drug loading and delivery materials [3, 5, 8, 9, 12, 14]. Moreover, there are various parameters that affect the morphology, diameter, distribution, and orientation of the produced textile nanofibers, namely the properties of the solution (e.g., viscosity, conductivity, and surface tension), the operating/process parameters (e.g., applied voltage, distance between the capillary tip and the collector, and the flow rate), and environmental conditions (e.g., humidity and temperature), and therefore, optimizing the variables that affect the electrospinning process is expected to develop textile nanofibrous materials with the specific features (porosity, wettability, mean diameter, and mechanical strength) desirable for tissue regeneration and wound-healing applications, as well as for PPE [3, 5, 8, 9, 12, 14].

Furthermore, the type of collector that is used in the electrospinning process determines the alignment of the electrospun nanofibers. Thus, using a stationary collector, the jet of the polymeric solution is directed to the collector, the solvent evaporates, and a dry and randomly oriented nanofibrous structure is obtained, mimicking ECM's three-dimensional structures [3, 5, 12, 14]. On the other hand, using rotating collectors (e.g., discs, rotating drums, and rotating wire drums), nanofibers are deposited with a specific orientation and, consequently, are more prone to medical textiles for nervous and muscle tissue regeneration. Also, when the collector rotation speed is increased, the nanofibers diameter reduces, and the alignment improves [3, 5, 12, 14].

Additionally, electrospun textiles have also been produced from a broad range of polymer materials. Among them, synthetic polymers approved by the US Food and

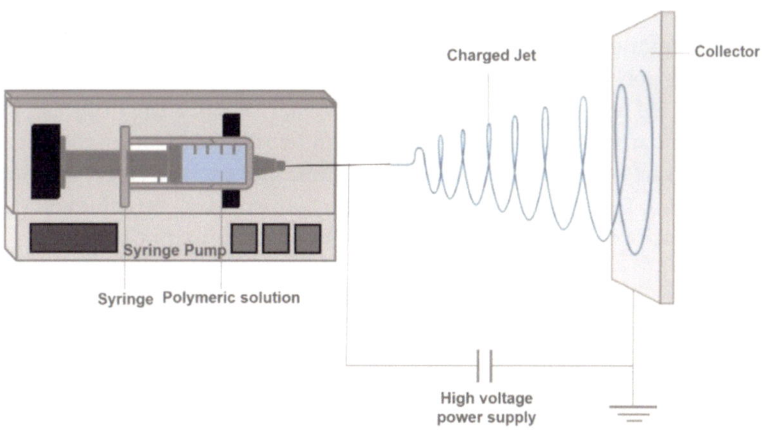

Figure 1.
Scheme of the basic device of electrospinning composed of four major parts: A spinneret, a syringe pump, a high-voltage source, and a collector.

Drug Administration (FDA), such as polycaprolactone (PCL), poly(L-lactic acid) (PLLA), polylactic-co-glycolic acid (PLGA), polyvinyl alcohol (PVA), polyethylene oxide (PEO), and polyurethane (PU) have been explored for drug loading and delivery and biomedical applications due to their biocompatibility, excellent thermal stability, mechanical properties, and proper biodegradation rates [5, 8, 10, 12, 16]. Nevertheless, more recently, researchers started to explore natural biopolymers, such as proteins (e.g., hyaluronic acid, collagen, and silk fibroin) and polysaccharides (e.g., chitosan, alginate, and gelatin) according to their inherent nontoxicity, biocompatibility, bioactivity, sustainability, and eco-friendliness, as well as blending of synthetic and natural polymers to design electrospun nanofibrous textiles with notable medical properties [5, 8, 10, 12, 16].

Besides, textile nanofibers have been successfully electrospun and loaded with different drugs, including water-soluble, water-insoluble, anticancer, and antibacterial drugs through distinct techniques (see Section 3) in order to protect these agents from decomposition within the body before reaching the target site and produce textile-based drug loading and delivery nanofibers with varied drug-release profiles and kinetics, such as biphasic release, prolonged release, immediate release, and targeted release [3, 12]. These strategies can also control the incorporation of multiple agents in order to achieve the desired therapeutic effect and reduce the number of dosing times, improving patient compliance [3, 11, 12]. In this way, the electrospun nanofibers produced from biodegradable and biocompatible polymers using different electrospinning techniques have received increasing attention as carrier materials for drug loading and delivery in medical fields due to their flexibility, effectiveness, and unique physicochemical properties. **Table 1** presents the different types of biocompatible synthetic and natural polymers that have been electrospun into nanofibers with controlled-release functions.

2.1 Mechanisms and kinetics for drug release

Textile nanofibers produced through the electrospinning process have been highlighted in the development of new nanomaterials, being able to provide improved drug-delivery systems with immediate- or prolonged-release profiles, minimum toxicity, and reduced dosage frequency due to their intrinsic properties (e.g., porosity, wettability, fiber diameters, and specific orientation), as described above [3, 5, 8, 10, 11, 17]. Besides, textile-based drug loading and delivery nanofibers fabricated from different types of polymer-carriers have been proposed to accurately predict the diverse drug-release kinetics from the polymeric matrix since the drug can interact with the polymer carriers in several manners and be released from nanofibers based on drug diffusion from pores, drug desorption from the surface layer, and/or polymer matrix degradation. Hence, different release mechanisms will be achieved depending on the characteristics of the drugs and the produced nanofibers (**Table 2**) [3, 5, 8, 11, 12, 17].

For example, occasionally, it is desirable that the incorporated drug has the ability to produce rapid effects, being released in an immediate form. In these cases, a faster rate of drug diffusion from polymer structure should be achieved by using water-soluble and biodegradable polymers, like PVA, PEO, and Polyvinylpyrrolidone (PVP), being largely influenced by several factors, such as polymer swelling, wettability, porosity structure, polymer erosion, and drug dissolution/diffusion and distribution [3, 5, 8, 11, 12, 16]. On the other hand, a prolonged drug release, also denoted as a

Natural polymers		
	Water-soluble	**Water-insoluble**
Proteins	—	Collagen Silk Fibroin Silk Sericin Keratin Zein
Polysaccharides	Alginate Gelatin Pectin Hyaluronic Acid Xylan Pullulan	Chitosan Cellulose Gum tragacanth
Synthetic polymers		
	Water-soluble	**Water-insoluble**
	Polyvinyl alcohol (PVA) Polyethylene oxide (PEO) Polyvinylpyrrolidone (PVP) Polyethylene glycol (PEG)	Polycaprolactone (PCL) Polylactic acid (PLA) Poly(L-Lactic acid) (PLLA) Poly(lactide-co-glycolide) (PLGA) Polyurethane (PU) Poly(methyl methacrylate) (PMMA) Polyglycerol Sebacate (PGS) Polystyrene (PS) Polyamide-6 Polyacrylonitrile (PAN) Cellulose Acetate Ethyecellulose

Table 1.
The most common biocompatible polymers used as drug-delivery carriers.

controlled, extended, and sustained release, requires the use of polymers or polymeric blends with a gradual degradation profile [3, 8, 11]. In addition, the electrospun nanofibers with an extended drug-release profile should display an adequate wettability and a thickness appropriate, as well as exhibit a core–shell structure containing a drug-loaded layer and an outer polymer layer that acts as a rate-controlling barrier. Moreover, core–shell nanofibers could provide biphasic drug-release profiles, which consist of an initial burst release followed by a sustained release (**Figure 2**) [3, 8, 11].

More recently, pulsatile drug-release profiles, where a sharp burst release is observed after a lag phase with no release, have gained increasing interest in various drugs or therapies. Besides, stimulus-responsive drug-release profiles have also been explored, and in these cases, responsive polymers after exposure to different stimuli, like pH, light, temperature, water, and CO_2, can change their physicochemical properties (**Figure 2**) [3, 5, 8, 9, 11].

Therefore, the solubility of the drug in the polymeric solution and the structure of the drug-loaded fibrous matrix can give valuable information about the release kinetics and consequently allow to tune of the desired behavior [3, 5, 8, 9, 11]. Concerning that, controlling the multiple electrospinning parameters that affect the nanofiber's features, such as fiber diameter, morphology, porosity, and wettability, and selecting the most suitable electrospinning technique for drug loading and subsequent

Drug-related factors	Effect on the drug-release kinetics
Drug loading amounts	Higher drug loading often results in faster release rates due to considerable amounts of surface-associated drug and the surface area-to-volume ratio of nanofibers.
Molecular weight of the drugs	Low molecular weight drugs display faster release rates.
Physical state of the drugs	Drugs in crystalline form are dispersed on nanofibers' surfaces, and a burst release is observed, while the drugs in the amorphous form are deposited deeper inside the nanofibers and released in a sustained manner.
Solubility of the drugs in the polymers used as delivery carriers	Generally, the drugs highly soluble in the polymer matrix are released slower.
Drug-Polymer interactions	Physical and chemical interactions between the drugs and the polymer matrix result in a slower release, while the direct incorporation of the drugs in the nanofibers is associated with an undesirable burst release.
Nanofibers-related factors	**Effect on the drug-release kinetics**
Molecular weight of the polymer	High molecular weight polymers exhibit slower drug-release rates from nanofibers.
Nanofibers' alignment	Randomly oriented fibers are often associated with quicker drug release due to the improved water uptake capacity.
Nanofibers' diameters	Higher fiber diameters result in extended-release rates since the drugs diffuse in a greater space before reaching the edge of the nanofibers. In turn, nanofibers with smaller diameters get a quicker release due to their higher surface area available and dissolution rate.
Nanofibers' crystallinity	Crystalline regions of the polymer are associated with a slower release when compared to amorphous domains.
Nanofibers' porosity	The high porosity exhibited by the nanofibers might increase liquid absorption and, therefore, result in a faster drug release. Nonetheless, the nanofibers' hydrophilicity, the pores sizes, and the volume of the pores influence the diffusion of the liquid which is absorbed by the nanofibers. Accordingly, it is frequently observed that thicker nanofibers with very high porosity release drugs faster in comparison to thinner fibers with low porosity.
Nanofibers' surface area-to-volume ratio	Nanofibers with a high surface area-to-volume ratio display a greater space for interact with the nearby liquid and, consequently, are observed to have a quicker drug release.
Electrospinning techniques for nanofibers production	The drugs can be loaded in different layers of nanofibers through different electrospinning techniques. The choice of the technique allows to control the drug location in the nanofibers and the drug-release kinetics from the polymers.

Table 2.
Effect of the drug and nanofibers-related factors on the release kinetics.

controlled release can produce good-quality textile-based drug loading and delivery materials [3, 5, 8, 9, 11]. In addition, the choice of the drug molecules and methodology applied to incorporate the drugs into the textile electrospun nanofibers can favor the location, amount, and timing of drug release needed to modulate the release profiles and achieve the desired effects [3, 5, 8, 9, 11].

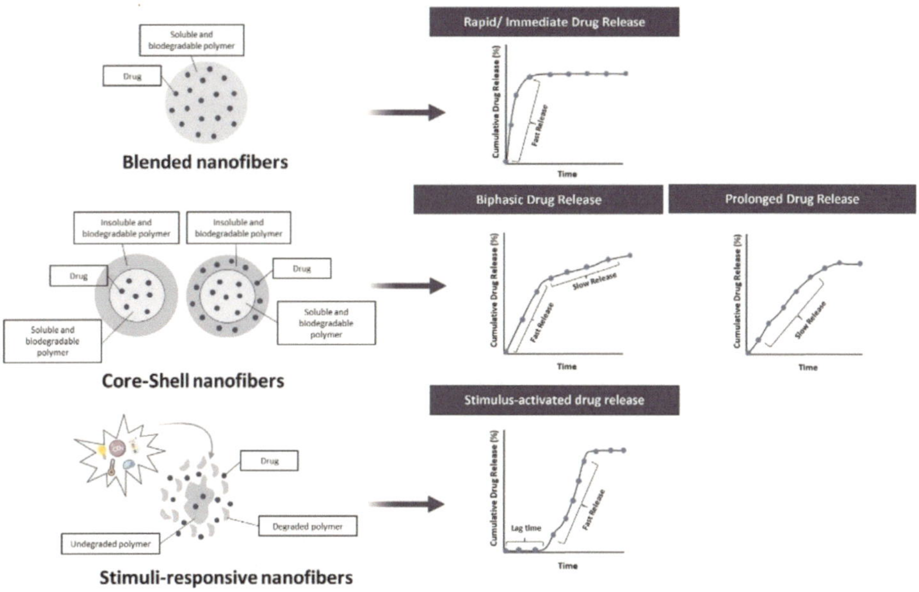

Figure 2.
Classification of the electrospun drug-loaded nanofibers based on their release profiles.

3. Electrospinning techniques for drug loading and delivery

A wide range of drug molecules, such as antibacterial molecules (e.g., antibiotics, nanoparticles (NPs), and natural extracts), molecules with biological activities, including anti-inflammatory, growth factors, vitamins, enzymes, and other proteins, as well as anticancer drugs, and nucleic acids (DNA and RNA) have been loaded into electrospun textile nanofibers through different electrospinning techniques, that comprise needle-based electrospinning and needleless electrospinning systems (**Figure 3**), in order to predict the location of drugs in the nanofibers and control the drug-release profile [3, 5, 8, 10–12, 16]. The drug loading using needle-based electrospinning techniques, like blending, side-by-side, coaxial, tri-axial, emulsion, and multi-needle electrospinning, add the drugs to the polymer solution before the electrospinning process and use a needle-like spinneret [3, 11, 12, 15, 18, 19]. On the other hand, needleless electrospinning methods, like the needle-free Nanospider technology, allow that multiple Taylor cones to be spun at the same time using a rotating electrode instead a needle, and hence this methodology is highly productive [3, 11, 12, 14, 18, 19]. In this way, the selection of the electrospinning technique can affect the role of the drug for its intended uses and manipulate its release rate and profile from the textile nanofibers [3, 11, 12, 14, 18, 19]. To accomplish that, the most common electrospinning methodologies for loading the drugs into electrospun textile nanofibers are described in detail below and their main advantages and disadvantages are summarized in **Table 3**.

3.1 Needle-based electrospinning

The needle-based electrospinning techniques would have a direct impact on the drug loading and release behavior from the electrospun nanofibers. The needle-based

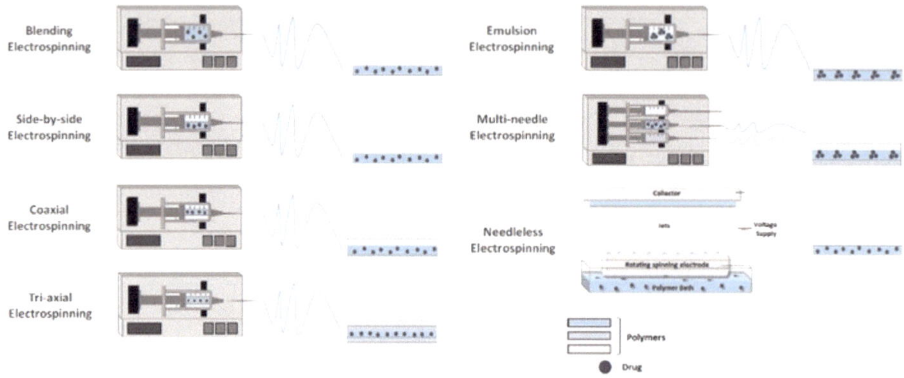

Figure 3.
Schematic representation of several electrospinning techniques for the preparation of drug-loaded nanofibers.

systems can result in simple, core-shell, and multi-axial nanofibers comprising more than one textile material [3, 11, 14, 18]. However, in the needle electrospinning techniques, capillary tubes, such as a needle-like spinneret, are required to form the nanofibers [3, 11, 14, 18].

3.1.1 Blending electrospinning

Blending electrospinning is the simplest and most basic method that uses a single nozzle to incorporate drugs into the electrospun nanofibers by blending them into the polymeric solution before the electrospinning process [14, 15]. In addition, different types of carrier materials and drugs can be processed by blending electrospinning, and hence it is the most researched and applied strategy for loading drugs into electrospun nanofibers. However, when this technique is used, in general, the drugs loaded are rapidly released from the nanofibers due to their homogeneous distribution all over the surface of the polymeric nanofibers and their high surface-to-volume ratio that favors the drug release. Nonetheless, their release depends on the degree of drug encapsulation into the polymeric matrix and the affinity between the drug and the polymer [14, 19]. Thus, to avoid an initial burst effect and ensure a more sustained release, the drugs absorbed and/or encapsulated into nanostructures, like NPs, nanospheres, nanomicelles, and nanotubes, can be added to the polymeric solution before the electrospinning process [14, 19].

Moreover, drugs with different properties can be blended into the same polymer carrier, and nanofibers with a biphasic drug release are obtained. For example, Li et al. dissolved PLGA (25% (w/v)) in a mixture of acetone and N,N-dimethyl formamide (DMF) with a volume ration of 3:1, and dexamethasone (DEX) at a concentration of 15% and green tea polyphenols (GTP) at concentrations of 5%, 10%, and 15% by weight of PLGA were added to the blend [22]. Afterward, the polymeric solutions were electrospun at a flow rate of 0.5 mL/h, using a working distance of 18 cm and an applied voltage of 23 kV. The results revealed that the DEX and GTP released from the electrospun nanofibers in PBS at pH 7.4 achieved a biphasic release profile. The hydrophilic GTP drug exhibited a faster release, while the hydrophobic DEX was successfully released from the channels formed by the fast-released GTP molecules. Thus, drugs with different hydrophilicity were used to achieve a biphasic release profile from the same polymer solution, which was controlled by the drugs' diffusion rate and the

Electrospinning Technique	Advantages	Disadvantages	References
Blending electrospinning 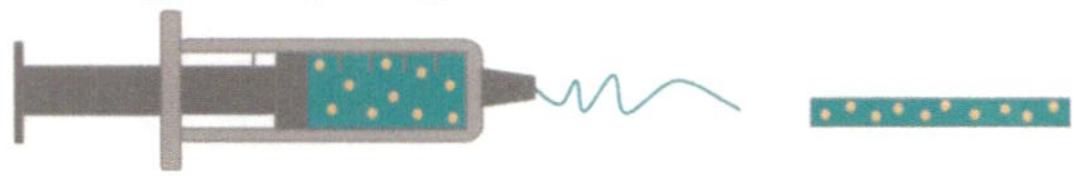	• The simplest and most basic electrospinning method. • A basic, single-nozzle electrospinning apparatus is required. • The properties of the materials can be adjusted by controlling the different electrospinning parameters.	• Only compatible polymers can be blended and used. • The drugs-loaded are rapidly released from the nanofibers.	[14, 19]
Side-by-side electrospinning	• Nanofibers with a Janus structure are produced. • A biphasic drug-release profile can be achieved. • Several polymers can be applied.	• Two capillaries are placed one adjacent to the other. • Both capillaries are controlled by the same pump and is applied the same voltage.	[3, 14, 18]
Coaxial electrospinning 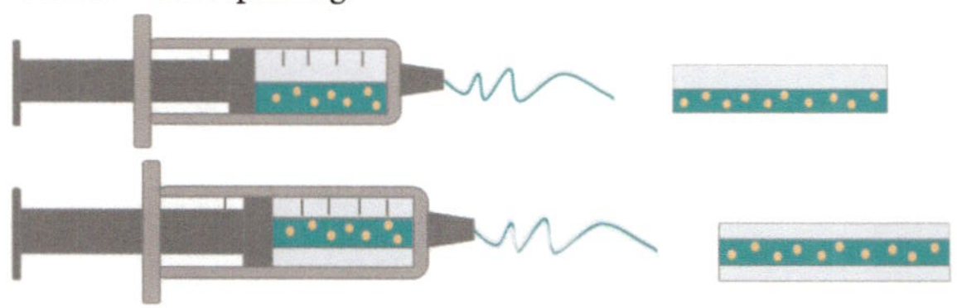	• Core-shell nanofibers can be fabricated with different core and shell compositions. • The core-shell nanofibers can be used as carriers for drug delivery. • Controlled drug-release profiles can be achieved. • Development of core-shell nanofibers with high drug-loading capacity. • Miscible and immiscible polymers can be used.	• Two concentric capillaries are used, and hence this technique requires a special apparatus. • The operating conditions should be carefully chosen to ensure desirable results.	[11, 14, 18, 19]
Tri-axial electrospinning 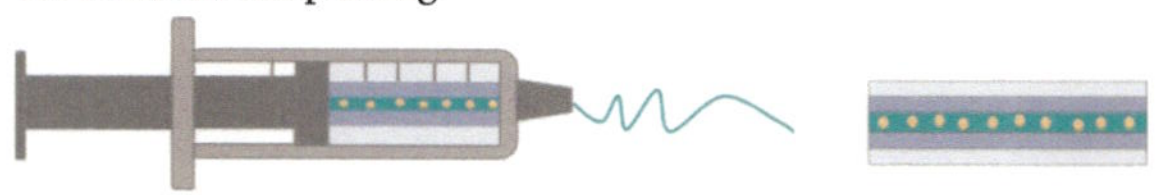	• Complex nanostructures can be formed through tri-axial electrospinning. • Potential to control drug loading and release, with possibility of loading multiple drugs. • Multiple drugs with different solubility can be used.	• Three concentric capillaries are required. • The electrospinning conditions should be carefully selected to ensure the required results.	[3, 14]

Electrospinning Technique	Advantages	Disadvantages	References
Emulsion electrospinning 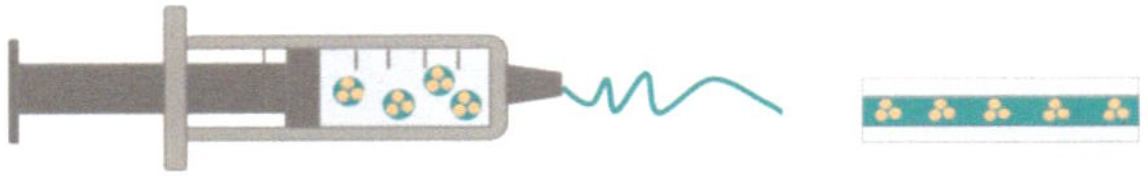	• A uniaxial electrospinning technique is used to produce core-shell nanofibers. • Diluted polymer solutions can be used. • A common solvent for both drug and polymer is not needed. • The drug-release mechanisms and kinetics can be controlled. • High drug-loading efficiency.	• Surfactants are required to stabilize the emulsions and the encapsulated drugs.	[11, 14, 18, 19]
Multi-needle electrospinning 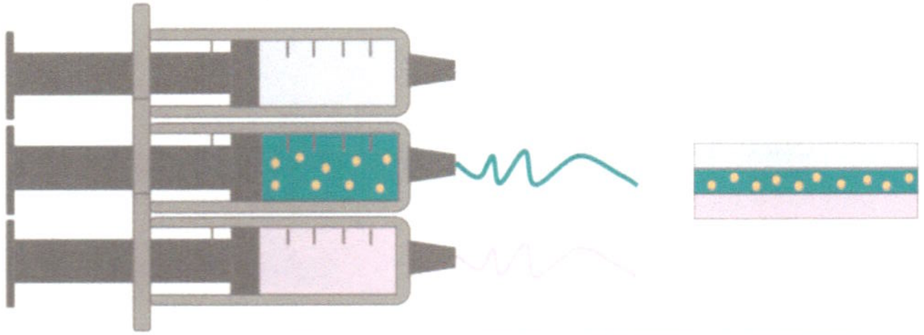	• Higher production rate of nanofibers. • Multiple drugs can be loaded. • Different release profiles can be achieved.	• Requires a special apparatus. • Unstable electric field strength. • Variation in fiber size distribution. • Blocking at the tip of the needles, and cleaning of multiple needle. • A higher voltage is required.	[20]
Needleless electrospinning 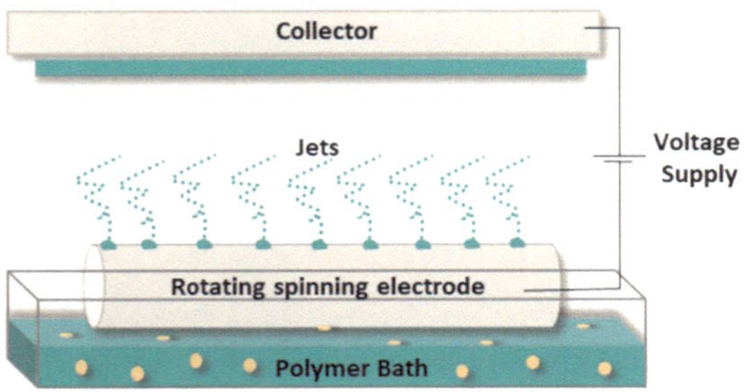	• Highly productive and more effective to produce high-quality nanofibers. • Lack of clogging.	• Requires a specific apparatus.	[18, 20, 21]

Table 3.
Advantages and disadvantages of the main electrospinning techniques used to produce textile-based drug loading and delivery materials.

degradation of the polymer carrier [22]. Furthermore, stimuli-responsive nanofibers could also be produced by blending electrospinning. Li et al. produced electrospun silk fibroin/PEO (SF/PEO) nanofibers with a pH-responsive controlled-release behavior [23]. For this purpose, 5 mol% strontium or copper-doped hollow bioactive glass nanospheres (5Sr/Cu-HBGNs), synthesized via a sol–gel method, were loaded with a 2.5 mg/mL aqueous solution of vancomycin hydrochloride (VAN), a glycopeptide antibiotic that inhibits bacterial cell wall synthesis, and incorporated in a mixture of 10% SF/3% PEO to produce 2% w/v solutions. The release kinetics behavior of VAN from VAN@5Sr/Cu-HBGNs/PEO/SF nanofibers in PBS at pH of 4.5 and 7.4 demonstrated that VAN@doped HBGNs/PEO/SF nanofibers presented a slower and sustained release rate in comparison with VAN@HBGNs/PEO/SF. Besides, the cumulative release of VAN from the electrospun VAN@5Sr/Cu-HBGNs/PEO/SF nanofibers at pH 7.4 was faster than at acidic microenvironments (pH 4.5), indicating that the drug release increased with an increase in pH along with a quick VAN desorption release. Thus, their pH sensitivity can be explained by the different solubility and both hydrophilic nature and amphoteric properties of SF under acid and alkali conditions [23].

3.1.2 Side-by-side electrospinning

The side-by-side electrospinning technique is a variation of the basic single nozzle electrospinning comprised of two capillaries placed one adjacent to the other to produce nanofibers with Janus structures and materials with two distinct layers [3, 14, 18]. However, this method is not easy to perform, since both capillaries, containing different polymer solutions, are controlled by the same pump, and the same voltage is applied to both solutions, which causes repulsion between the two different polymers and separate from each other [3, 14, 18]. Hence, few studies have been developed in order to fabricate electrospun textile nanofibers with a Janus structure [3]. However, the side-by-side bi-component fibers exhibit the characteristics of both polymers, and consequently, we can achieve a biphasic drug-release profile required for a particular type of drug and application [14]. Thus, a biphasic drug release can be achieved by incorporating the drug on both sides of the fiber, where one side consists of a water-soluble polymer and the other side of a water-insoluble polymer [14]. On the side composed of the water-soluble polymer, the drug diffusion rate will be affected by polymer erosion and will be faster, while on the side containing the insoluble polymer, the drug release will be extended [14].

Shi et al. used side-by-side electrospinning to combine a solution of 10% (w/v) PLGA with an anti-inflammatory drug (0.2% (w/v) valsartan, V) and a solution of 15% (w/v) PVA with copper sulfide nanoparticles (CuS NPs) and an antibacterial drug (mupirocin, M) [24]. Both solutions were placed in two different syringes with 26 G blunted spinning needles and electrospun at a constant feeding rate of 1.0 mL/h, using a working distance of 20 cm and an applied voltage of 15 kV. The Janus amphiphilic nanofibers composed by a hydrophobic side of PLGAV-CuS NPs and a hydrophilic side of poly(vinyl alcohol) mupirocin nanofiber (PVAM) exhibited a biphasic drug release mechanism. The cumulative release of the hydrophilic antimicrobial drug, mupirocin, from PLGAV-CuS/PVAM nanofibers enabled a continuous and slow release, reaching 89.36% after 24 h of incubation [24]. The sustained release of mupirocin can be attributed to the hydrogen bonding between the drug and the PLGAV-CuS/PVAM. In addition, the water uptake ability of the PLGAV-CuS/PVAM nanofibers could have a certain blocking effect on the release of mupirocin. In turn, the release profile of valsartan reached only 23.77% after 24 h, providing a slow release of drug from nanofibers. This result confirms that the diffusion of hydrophobic drugs in a hydrophilic environment is hindered. Therefore, the PLGAV-CuS/PVAM Janus nanofibers initially released mupirocin more quickly, and then, the valsartan and remaining mupirocin were continuously released, achieving tunable antibacterial and anti-inflammatory gradient drug release systems [24].

3.1.3 Coaxial electrospinning

Coaxial electrospinning technique also uses two capillaries, like side-by-side electrospinning, however, in coaxial electrospinning the capillaries are arranged one inside the other. In addition, the concentric capillaries, i.e. the inner capillary coaxially placed inside the outer one, are connected to two independent reservoirs, with controllable flow rates [11, 14, 18]. Concerning that, coaxial electrospinning allows the production of core-shell nanofibers with different core and shell compositions, including functional fibers encapsulated with different drugs and hollow nanofibers with adjustable shell thickness by selectively removing the core from as-spun core-shell nanofibers [11, 14, 18].

Hence, core-shell nanofibers produced from coaxial electrospinning provide an added advantage as carriers for drug delivery, protecting the drugs' native structure and their bioactivity from harsh environments during textile nanofibers production [11, 14, 19]. In addition, the drugs loading into the core of the textile nanofibers can significantly reduce and/or prevent the initial burst release rate from the polymeric matrix and maintain a sustainable drug release for an extended and controlled time [11, 14, 19]. Moreover, this technique contributes greatly to the high loading capacity of diverse drugs and allows susceptible drugs to be delivered. Furthermore, core-shell nanofibers prepared from miscible and immiscible polymers can be produced through coaxial electrospinning, as well as from nonspinnable solutions and polymers and drugs with low compatibility [11, 14, 19]. However, this technique requires a special apparatus, and the operating conditions should be carefully chosen to ensure desirable results [11, 14]. Nevertheless, coaxial electrospinning has the ability to adjust the thickness of both layers *by changing the flow rate*, particularly the shell thickness that acts as a barrier to diffusion, controls the release kinetics of the encapsulated drugs, and enables the potential to simultaneously load various components [14, 19]. Therefore, core-shell nanofibers fabricated from the coaxial electrospinning technique can effectively avoid the initial drugs' explosive release and achieve a sustained release profile. For example, Li et al. successfully produced Xylan/PCL core–shell nanofibers using coaxial electrospinning. In this study, Xylan aqueous solution (20 *w/w*%) containing the anti-inflammatory drug levofloxacin (LEV) was used as core and coated by (12 *w/w*%) PCL solution (shell). The solutions were placed in two coaxial injectors and electrospun at an applied voltage of 14 kV, using a collecting distance of 15 cm, a constant flow rate of 1.0 mL/h for the Xylan solution and a flow ration between Xylan and PCL of 1:1.1, 1:1.2, and 1.1.4, respectively. The results obtained revealed that the hydrophobicity of the PCL prevents a burst release and short diffusion of the LEV loaded into the water-soluble Xylan core. Additionally, the increase in PCL's flow rate resulted in a lower degradation rate of the nanofibers due to the thicker fibrous shell layer and consequently can be used to produce a slow drug release [25].

3.1.3.1 Tri-axial electrospinning

Coaxial electrospinning can be used to produce textile nanofibers with more complex architectures, namely by using a tri-axial electrospinning technique, composed of three concentric capillaries [3, 14]. The three-layer core-shell nanofibers are formed from an external polymeric solution, an intermediate polymeric solution, and a core polymeric solution controlled by different pumps [3, 14]. Besides, the tri-axial nanofibers can be produced with the core enclosed by two different polymer layers or with two layers of the same polymer. Moreover, the core polymers can also be surrounded by an outer polymer layer and a layer of void areas. In turn, different drugs can be incorporated into the core, or in each different layer, and released by diffusion and degradation [3, 14]. The three-layer textile nanofibers can be also prepared with targeted drug-release functions for medical applications in order to improve the drug efficacy and reduce its toxicity and side effects, as well as with a linear and constant release rate in order to eliminate the drug burst release [3, 14]. Likewise, the drug content can be changed by using a tri-axial electrospinning technique, creating an increased drug gradient distribution from the outer layer to the inner layer. Thus, in comparison with the double-layer core-shell nanofibers produced from coaxial electrospinning, the tri-layer core-shell nanofibers prepared by tri-axial electrospinning provide more potential to overcome the limitation of poor drug solubility, protect drugs from the adverse environments, and control the drug-release kinetics for developing functional textile-based drug loading and delivery materials [3, 14, 26].

Ding et al. compared the capacity of aspirin-loaded ES100 core-shell nanofibers (produced with an improved tri-axial electrospinning system) (CSFs) with the monolithic composite (prepared with the same material using a traditional blended single-fluid electrospinning) (MCFs) for a sustained drug release. The CSFs revealed better results than the MCFs, although the release of aspirin from both occurred through an erosion mechanism [27]. The CSFs exhibited a prolonged release of the drug and showed a lower release rate during the first 2 h under acidic conditions. In addition, in a posterior neutral environment, a prolonged drug release was obtained by an extended-release effect [27].

3.1.4 Emulsion electrospinning

Emulsion electrospinning is a uniaxial electrospinning technique, which has attracted growing interest in the production of core-shell textile nanofibers without the use of specific coaxial apparatus [11, 14, 18, 19]. This method is similar to conventional electrospinning, apart from that the polymeric solution is replaced by a water-in-oil (W/O) or oil-in-water (O/W) emulsion [14]. In the emulsion electrospinning, most emulsions are of the W/O type, where the drug is usually dissolved in an aqueous phase composed mainly of water-soluble polymers (i.e., the water phase) and then dispersed in a continuous-phase composed of an organic polymer solution (i.e., the oil phase) [14, 19]. The droplets of the water phase dispersed in the oil phase evaporate more slowly than the organic polymer solution, resulting in a viscosity gradient between the two different phases. Subsequently, this gradient guides the droplets of the water phase from the surface to the center, the droplets are stretched into elliptical shapes along the axial region under a high voltage and the core-shell nanofibers containing the drug into the core are formed [14]. Hence, W/O emulsions, in which the hydrophilic drugs are dissolved in an aqueous polymer solution, while hydrophobic polymers are dissolved in organic solvents, are particularly useful for a drug continuous and sustained release, avoiding the initial burst release, typical of most drug-delivery materials, as well as play a role in protecting the drugs in the core from the harmful effects of the external environment, enhancing their bioactivity and effectiveness [11, 14, 19]. Moreover, the drugs are released faster in the shell than in the core layer; once the drugs are loaded into the core have to pass through the core-shell matrix before being released. Furthermore, good-quality core–shell nanofibers can be produced through emulsion electrospinning using diluted polymer solutions [19]. Various combinations of hydrophilic drugs and hydrophobic polymers with low compatibility and affinity can also be explored, removing the necessity of using a common solvent for both the drug and the polymer [18]. However, emulsifiers such as surfactants are frequently used to stabilize the emulsions and the encapsulated drugs [11, 19].

Weng et al. prepared camellia oil-loaded zein nanofibers with a core-shell structure using an emulsion electrospinning technique. For this purpose, zein (30% *v/v*) aqueous solution was firstly prepared in acetic acid (70% *v/v*) and then camelia oil was dropped at 10, 20, 40, and 60% (*v/v*) based on the zein solution. After stirring for 1 h, 30% (*w/w*) of glycerol (based on the weight of zein) was added as a plasticizer. The blends were mixed with a high-speed homogenizer at 1000 rpm for 3 min and then treated by ultrasonic for 2 min at 250 W at 25°C to obtain the required O/W emulsions. Afterward, the emulsions were placed in the syringe and electrospun at a flow rate of 0.5 mL/h, using a working distance of 12 cm and an applied voltage of 20 kV. The results of this study demonstrated that core–shell nanofibers of zein with camellia oil could be used as a promising controlled-release carrier for hydrophobic bioactive compounds [28].

3.1.5 Multi-needle electrospinning

The multi-needle electrospinning is one of the simplest techniques to achieve nanofibers with high throughput [20]. This electrospinning technique implies passing a polymeric solution through multiple needles connected to a high-voltage supply. In addition, the flow rate is controlled by a single pump and is required the application of a higher voltage to continuous electrospinning due to the large mass of the spinning solution [20]. Nevertheless, multi-needle electrospinning presents several drawbacks, such as unstable electric field strength, changes in fiber size distribution, clogging at the tip of the needles, and cleaning of multiple needles. In addition, the multi-needles may also provide repulsion from adjacent jets. Nonetheless, multiple drugs can be added, and diverse release profiles can be achieved [20].

Varesano et al. tested several multi-jet electrospinning setups using a 7 wt% PEO aqueous solution and varied the number of nozzles between 2 and 16. The data obtained showed that when increasing the number of jets above 6 was imperative to increase the collector dimensions because the deposition area also increased [29]. Hence, multi-needle electrospinning systems have great potential for large-scale nanofibers production. Moreover, Yoon et al. fabricated Polystyrene (PS)/Polyamide 6 (PA6) nanofibers using multi-jet electrospinning and revealed that the fiber content was directly related to the syringes' number used for PS and PA6, respectively [30].

3.2 Needleless electrospinning techniques

The needleless electrospinning does not use a needle-like spinneret but static or rotating spinnerets, which can be a cylinder, a ball, a disc, and/or a rotating electrode, immersed into an open container with the polymeric solution in order to produce nanofibers in higher quantities for industrial range [18, 20]. Nevertheless, when the open container is filled with a polymer solution dissolved in volatile solvents, the solvent evaporates quickly and can affect the reproducibility of fiber morphology. Moreover, in needleless electrospinning, many Taylor cones (the source of nanofibers) are created simultaneously on the surface of the spinnerets, and hence these techniques are highly productive and more effective in producing high-quality nanofibers [20, 21]. For example, in the Nanospider technology, a needleless electrospinning technique, a rotating electrode/roller transport the polymer solution from the container, allowing industrial-scale production of nanofibers [20, 21]. In this way, Nanospider is a pilot electrospinning equipment, already used on an industrial scale, which allows an easy scale-up. Furthermore, this is a simple and versatile technique for textile nanofibers production from a wide range of materials [20, 21]. Regarding that, Mouro et al. incorporated *Hypericum perforatum* L., a medicinal plant extract, in a PLLA/PVA/CS emulsion via Nanospider, a needleless electrospinning technology. The results obtained showed a controlled and sustained release profile of the *Hypericum perforatum* L. from the produced nanofibers for 72 h. In addition, the release speed was dependent of the content of plant extract incorporated and the swelling of the produced nanofibers [31].

Recently, electrospinning derivative technologies, such as melt electrospinning and centrifugal electrospinning, have also arisen as promising approaches [3, 8, 14, 32, 33]. The melt electrospinning is a solvent-free process and hence a more efficient and environmentally friendly technique. A typical melt electrospinning setup includes a heated polymer fed into a spinneret. However, like traditional electrospinning, a high electric potential is applied between the spinneret and the collector for producing the textile nanofibers [8, 14, 33]. Moreover, melt electrospinning enables the production of nanofibers in the absence of residual solvents because it does not require the dissolution of the polymers in organic solvents. In addition, various materials can be used to fabricate nanofibers. However, this technique only uses materials that melt, and polymer degradation or denaturation can occur because high temperatures are applied [8, 14, 33]. On the other hand, the centrifugal electrospinning technique combines the electrostatic force of the electrospinning and the centrifugal force to promote jet stretching [3, 32]. Centrifugal electrospinning is a simple and cost-effective method that allow the fabrication of highly aligned and small-diameter textile nanofibers, upon applying adjusted rotational speeds and voltages, for drug delivery and regenerative medicine applications [3, 32]. Regarding that, in centrifugal electrospinning, the electrical fields allow the spinneret to rotate at a slower speed and simultaneously a lower voltage to overcome the surface tension of the spun solution is required [3, 32].

4. Applications of the electrospun textile drug-delivery materials in medical fields

The unique features of the electrospun textile materials, namely the high surface area, tunable porosity, and smaller diameter of the fibers, as well as the easy optimization of the nanofiber's properties through controlling the different parameters, like the solution properties, processing parameters, and environmental conditions, made them ideal for drug loading and delivery applications [3, 5, 8, 9, 12, 14]. Moreover, the various electrospinning techniques allow for incorporating a wide range of drugs with high loading and encapsulation efficiencies into diverse polymer-textile nanofibers [3, 5, 11, 12, 14, 18, 19].

Therefore, this section is focused on the textile-based drug loading and delivery of nanofibrous materials produced from the different electrospinning techniques for medical applications. Regarding that, it is provided an overview of the recent studies for applications in bone, nerve, periodontal, and vascular regeneration, as well as in wound dressings and other textile-based materials, like PPE and cancer treatment. Furthermore, **Table 4** are presented several studies where the authors prepared drug-loaded nanofibers using different electrospinning technologies for medical applications.

4.1 Bone tissue regeneration

Nanofibers produced from electrospinning have been investigated for bone regeneration due to their intrinsic properties, such as biocompatibility, highly porous structure, small pore size, high surface area, three-dimensional (3D) architecture, and mechanical features that are compatible with the natural nanostructure of bone. In addition, different drugs have been incorporated into these nanofibrous structures to improve bone regeneration [8, 16].

In 2022, Canales et al. produced electrospun fibers of Poly(lactic acid) (PLA) containing bioactive glass (n-BG) and magnesium oxide (n-MgO) NPs to be used in bone tissue engineering applications [49]. Their results showed that the electrospun PLA/n-BG and PLA/n-BG/n-MgO fibers presented a significant increase in fiber diameter with mean diameter values of 3.1 ± 0.8 μm, while the neat PLA and the PLA/n-MgO displayed an average fiber diameter of 1.7 ± 0.6 μm [49]. Besides, the electrospun PLA/n-BG/n-MgO fibers presented a high porosity and an interconnected pore structure required for tissue engineering applications. However, the incorporation of the NPs affected the thermal properties and mechanical properties [49]. All the composite fibers containing n-BG showed the capability to precipitate hydroxyapatite on the surface, demonstrating their bioactivity, while the addition of the n-MgO to PLA provides antibacterial properties against *S. aureus* [49]. Moreover, the osteoblastic phenotype expression ability of the produced composite fibers was assessed in comparison with the neat PLA fibers measuring the alkaline phosphatase expression (ALP), a marker of osteoblastic activity known for its capacity to promote cell differentiation, and the results indicated that the PLA/n-BG presenting the highest osteoblastic expression. Therefore, this study revealed that the incorporation of both n-BG and n-MgO NPs into the PLA can be considered to produce a synergic effect increasing its bioactivity and antimicrobial behavior [49].

Huang et al. encapsulated citrate-stabilized gold-nanoparticles (GNPs) into PVP/Ethylcellulose (EC) by coaxial electrospinning technique in order to create core-shell nanofibrous membranes [50]. In this study, the core-shell nanofibers were prepared by coaxial electrospinning with EC as the core and PVP containing GNPs as the shell. The GNPs-loaded electrospun PVP/EC nanofibers were prepared by changing the feeding GNPs to 0.5:1, 1:1, and 1.5:1 (P/E-0.5, P/E-1, and P/E-1.5), respectively [50]. The data obtained revealed that the GNPs were successfully encapsulated into the electrospun nanofibers. Besides, the GNPs addition did not significantly affect the morphology of the nanofibers, although it improved the porosity and the mechanical properties of the nanofibers [50]. Furthermore, the presence of the GNPs in electrospun PVP/EC membranes triggered a higher alkaline phosphatase activity, mineralized nodule formation, and osteogenic-related genes expression, thus confirming the excellent biocompatibility and osteogenic bioactivities of the produced composite material [50]. The *in vivo* studies also confirmed that GNP-incorporated electrospun PVP/EC nanofibers accelerate bone regeneration. Hence, the produced coaxial electrospun membranes demonstrated the appropriateness for being

Medical application	Drugs/biological agents	Carrier/polymer	Drug-release highlights	Reference
Wound treatment	Amoxicillin (AMX)	Pullulan (Pull) and poly(lactide-co-glycolide) (PLGA)	The release rate of the drug from AMX-loaded Pull nanofibers was fast. In turn, when the AMX-loaded Pull nanofibers were sandwiched between extra electrospun PLGA layers was observed a continuous sustained release favorable for wound-healing dressings. The PLGA layers around the AMX-loaded Pull work as a diffusion barrier, decreasing the release rate of AMX.	[34]
	Ceftazidime (CTZ)	Polycaprolactone (PCL) and Gelatin	The sandwich-like PCL/Gelatin/PCL electrospun multilayered mats were incorporated with CTZ. The release profile of CTZ-loaded Gelatin from the internal layer was found to be dependent on the PCL structure, which stabilized and trap the CTZ, avoiding a fast initial burst. The result was a continuous and slower sustained release profile.	[35]
	Tannic acid (TA)	Polyamide-6 (PA6) and Hydroxyethyl cellulose (HEC)	Three different ratios of TA (5, 10, and 25% wt, versus the total weight of polymer) were incorporated into the polymer blend solution. In vitro release of TA from PA6/HEC nanofibers displayed no burst release. The PA6/HEC/25%TA nanofibers exhibited a stable TA release rate that reached 80% after 24 h of incubation. Thus, the release rate has a direct relation with the TA concentration (up to 25%).	[36]
	Thrombin (TMB) and Vancomycin (VCM)	Polyvinyl alcohol (PVA), Gelatin, and PLGA	The electrospun nanofibers loaded with TMB, a hemostatic agent, and VCM, an antibacterial agent, were placed in PBS pH 7.4 for 72 h and revealed a burst release	[37]

Medical application	Drugs/biological agents	Carrier/polymer	Drug-release highlights	Reference
			within the first 2 h followed by a sustained release, indicating its potential as a dressing material for hemostatic wound-healing applications.	
	Naringenin (Nrg)	PCL and Polyethylene glycol (PEG)	The Nrg was loaded into PCL/PEG nanofibers with various amounts (3.125, 6.25, 12.5% Wt. Nrg to polymer) and the release profiles revealed a burst at the first 2 h. Then, a steady release was observed, reaching approximately 70, 42, and 19% within 24 h for PCL/PEG/Nrg (3.125%), PCL/PEG/Nrg (6.25%), and PCL/PEG/Nrg (12.5%) NFs, respectively. So, interestingly the release rate was faster in the nanofibers with low Nrg ratio.	[38]
Nerve regeneration	Nimodipine	PLGA	1 and 10% of Nimodipine, a calcium-antagonistic drug with preferential cerebrovascular action and neuroprotective properties, was successfully loaded into PLGA nanofibers. The results indicated a controlled and sustained drug release over 4–8 days, highly dependent on release conditions. Therefore, the electrospun Nimodipine-loaded PLGA nanofibers are a promising approach for intracranial use.	[39]
	Melatonin (MLT) and Fe_3O_4 magnetic nanoparticles (Fe3O4-MNPs)	PCL	The multilayered composite loaded with MLT and Fe3O4-MNPs showed a timely release of MLT suitable for peripheral nerve repair.	[40]
Bone repair	Cephalexin Monohydrate (CEM) and Silver (Ag)	Silk fibroin (SF), Mesoporous silica (MSi), PVA	The SF containing the CEM drug was embedded into PVA with Ag doped MSi NPs to produce three-	[41]

Medical application	Drugs/biological agents	Carrier/polymer	Drug-release highlights	Reference
			dimensional sandwiched nanofibers. The prepared sandwich layered nanofibers exhibited a synergistic antibacterial effect due to the controlled and sustained release of CEM from the middle layer. No burst release was observed.	
	Vancomycin hydrochloride (Van), Simvastatin (Sim), and Fluorescein isothiocyanate-bovine serum albumin (FITC-BSA)	PCL and Gelatin	PCL/Gelatin co-electrospinning membranes were loaded with different model drugs (*e.g.* Van and FITC-BSA as hydrophilic drugs/proteins and Sim as a hydrophobic drug) in order to evaluate their potential for bone regeneration. The results showed that the presence of the PCL can avoid a fast degradation and dissolution of the Gelatin and consequently an early explosive release of the hydrophilic drugs/proteins. Thus, by increasing the PCL content extended drug release can be obtained.	[42]
Tendon healing	Growth factors: PDGF-BB, FGF-2, and TGF-β1	PCL, Polyethylene oxide (PEO), and Hyaluronic acid (HA)	Core-shell nanofiber membranes were produced via coaxial electrospinning to promote tendon recovery posttendon injury. The nanofibers comprised a PCL shell and a core of PEO and different ratios of HA/Platelet-rich plasma (PRP) where were loaded different growth factors. The results showed that the release of growth factors from the core region of the core-shell nanofibers was linear for 10 days and followed a sustained release behavior after 35 days.	[43]
Eye disease	5-Fluorouracil (5-FU)	PVA, *Trigonella foenum-graecum*(TFG),	Electrospun composite nanofibers composed of PVA, TFG, and CS were	[44]

Medical application	Drugs/biological agents	Carrier/polymer	Drug-release highlights	Reference
		and *Coriandrum sativum* (CS)	incorporated with 5-FU, a water-soluble, hydrophilic antimetabolite drug. The composite nanofibers showed a sustained drug release over 4 h, emphasizing their suitability for ocular drug-delivery systems.	
Cancer treatment	Paclitaxel	PCL and Chitosan	The paclitaxel-loaded liposome-incorporated Chitosan/PCL-Chitosan core-shell nanofibers (produced with a coaxial electrospinning system) do no demonstrate an initial burst release under Ph values of 5.5 and 7.4. Additionally, a linear release of paclitaxel from the liposome-loaded core–shell nanofibers was observed, making them effective for the treatment of various tumors.	[45]
	Docetaxel (DTX)	Gelatin and Polylactic acid (PLA)	Core–shell Gelatin/PLA nanofibers containing an anticancer DTX drug were obtained through emulsion electrospinning. The release profiles revealed that the cumulative DTX release was higher in acidic pH (6.5) than in neutral pH (7.4), enhancing the efficacy of drug delivery to tumor tissues in an acidic environment.	[46]
	Doxorrubicina (DOX)	PCL and Gelatin	PCL/Gelatin nanofibers containing DOX-loaded amine-functionalized mesoporous silica NPs (MSNs), linked with green synthesized Ag NPs, revealed a proper release profile of DOX and Ag NPs for healing wounds caused by melanoma cancer.	[47]
Vascular regeneration	microRNA-126 and microRNA-145	PCL, poly (ethylene glycol)-b-poly(l-lactide-co-ε-	A trilayered material was prepared with a PCL outer layer and encapsulated with microRNA-126 and	[48]

Medical application	Drugs/biological agents	Carrier/polymer	Drug-release highlights	Reference
		caprolactone) (PELCL)	microRNA-145 in the Arg-Glu-Asp-Val (REDV) peptide-modified PELCL inner and PELCL middle layers, respectively. The data obtained showed a fast release of microRNA-126 and/or a slow release of microRNA-145 suitable for replacement of artificial small-caliber blood vessels.	

Table 4.
Several drug-loaded nanofibers using different electrospinning techniques for medical applications.

considered as a promising material for bone repair [50]. In a similar way, Alam et al. produced coaxial PCL/Gelatin/Poloxamer 188 (P-188) nanofibers for duel release of β-lactoglobulin (a hydrophilic protein) and vitamin K2 (a hydrophobic agent) from core and shell fibers [51]. The coaxial electrospun nanofibers produced from the PCL/P-188 shell and Gelatin/P-188 core displayed superior stability, average weight loss, no water uptake ability, and a sustained, controlled release of both β-lactoglobulin and vitamin K2 [51]. Moreover, these nanofibers also significantly enhanced alkaline phosphatase activity and promoted higher cell viability for human osteogenic sarcoma cells (Saos-2 cells), and thereby, coaxial PCL/Gelatin/P-188 nanofibers can release both hydrophilic and hydrophobic bioactive compounds and be a promising bioactive material for bone tissue engineering with improved osteogenic properties [51].

Electrospun nanofibers with a core-shell structure have also been produced by emulsion electrospinning for bone tissue regeneration. For example, Boraei et al. fabricated core-shell nanofibers from PVA and PCL by water-in-oil electrospinning encapsulated with strontium ranelate (SrR), an osteogenic agent [52]. The authors successfully obtained nanofibers with a shell of PCL (the oil phase) and a core of PVA incorporated with SrR (the water phase). In addition, higher contents of SrR induced the formation of fibers with increased diameters and decreased the crystallinity of the nanofibers [52]. Moreover, the SrR release from the electrospun core-shell nanofibers occurred through a Fickian diffusion mechanism, and it was more evident a quicker drug dissolution in the samples with higher SrR content [52]. Additionally, the incorporation of the SrR improved mesenchymal stem cell proliferation and enhanced the expression of ALP, Runx2, Col I, and OCN genes. The *in vivo* assays also showed that animal models treated with the core-shell nanofibers displayed an increased bone formation of calvarial defects [52]. Further, Al-Baadani et al. investigated the suitability of the electrospun PCL/Gelatin membranes containing different drugs/proteins, like fluorescein isothiocyanate-bovine serum albumin (FITC-BSA), vancomycin hydrochloride (Van), and simvastatin (Sim) to improve the antibacterial and osteogenic activities [42]. This approach combined the biocompatibility of the Gelatin, namely its ability to improve the adhesion and differentiation of osteoblasts, with the mechanical properties of the PCL. The results showed that the produced membranes from coaxial electrospinning could be utilized as drug-carriers for slow-release hydrophobic drugs, like Sim, loading in the PCL solution, and for controlled-release of hydrophilic drugs/proteins, like FITC-BSA and Van, loading in Gelatin solution, adjusting the PCL content [42].

Additionally, the electrospun nanofibers have also been applied to restore other structural tissues, namely skeletal muscle and cartilage ligament.

4.2 Nerve tissue regeneration

Neural tissue repair is one of the major challenges faced by the researchers because the most common neural injuries result in an irreversible loss of function [16]. In this context, biocompatible polymer nanofibrous conduits, with customized lengths and sizes and controlled delivery of drugs for nerve regeneration, have recently been investigated using different electrospinning techniques. In addition, both aligned and randomly oriented nanofibers have been further explored in order to improve neural stem cell adhesion and differentiation, as well as neural tissue regeneration [8, 16].

Alipour et al. explored the encapsulation of fingolimod-loaded PLGA nanoparticles into electrospun PU/PCL/Gelatin nanofibers for promoting neurite formation and axonal regeneration. Different amounts of fingolimod (0.01, 0.02, and 0.03%), a useful drug in nerve regeneration, were studied. The increased content of drug loaded into the membranes resulted in an increase in the nanofiber's mean diameter [53]. Nevertheless, the nanofiber's diameters of the produced membranes are in the acceptable range displayed by nerve tissues, as well as the degradation rate, the mechanical properties, and the water uptake behavior [53]. Besides, the electrospun PU/PCL/Gelatin nanofibrous membrane loaded with 0.01% fingolimod through PLGA NPs provided a more favorable environment for cell growth and proliferation. Moreover, the release profile of the fingolimod from this membrane displayed a burst release behavior during the first 24 h, and then a slow release occurred for up to 5 days [53]. Therefore, the electrospun PU/PCL/Gelatin nanofibers loaded with fingolimod NPs can be a suitable candidate for application in neural tissue engineering [53].

Xia et al. incorporated the recombinant human vascular endothelial growth factor (VEGF) and the recombinant human nerve growth factor (NGF) on the shell and in the core of the PLLA nanofibers by emulsion electrospinning [54]. The electrospun core-shell nanofibrous membranes with VEGF on the shell and NGF in the core displayed a sequential release profile, in which most of the VEGF was released in the first few days, and the NGF could be gradually released from the PLLA core nanofibers for >1 month [54]. Besides, the nanofibers containing the NGF and VEGF encouraged the nerve differentiation of induced pluripotent stem cells-derived neural crest stem cells (iPSCs-NCSCs) in vitro [54]. Moreover, in vivo data showed that, after 3 months, animals treated with the dual-delivery of VEGF and NGF membrane showed a significant improvement in neovascularization and nerve healing postoperation [54].

In turn, Alipour et al. reported the incorporation of vitamin C (VC) at concentrations of 5, 10, and 15 wt.% into PCL/Polyglycerol Sebacate (PGS) nanofibers [55]. The nanofibers produced from a rotating drum collector presented an aligned and uniform appearance, although the incorporation of the VC induced a decrease in the nanofibers' mean diameter. In addition, the electrospun membranes exhibited appropriate properties for nerve applications, namely mechanical properties, wettability, water uptake capability, and degradation [55]. Furthermore, the release profile of VC was characterized by an initial burst, followed by a gradual release, which is fundamental for more beneficial therapeutic effects on peripheral nerve regeneration. However, the electrospun PCL/PGS nanofibers containing 5 wt.% VC provided a more favorable environment for PC12 cell adhesion and migration [55].

Chen et al. prepared multilayered membranes with a PCL outer layer, a Fe_3O_4 magnetic NPs (Fe_3O_4-MNPs)/PCL middle layer, and a melatonin (MLT)/PCL inner layer for peripheral nerve regeneration [40]. The results obtained revealed a sequential and sustainable drug release appropriate for mitigating oxidative stress and inflammatory response and inducing a microenvironment suitable for nerve regeneration [40]. In fact, the composite nanofibrous membranes showed satisfactory mechanical strength, in vitro biocompatibility in contact with rat Schwann cells (RSC 96), and beneficial effects for nerve regeneration in vivo [40]. Further, Mohamadi et al. obtained electrospun conduits from the PCL/collagen/nanobioglasses (NBG) blends, and the NGF was loaded in the structure of conduits [56]. Their results showed that the PCL/Collagen/NBG containing NGF revealed the potential to be applied to promote sciatic nerve regeneration. In addition, the NGF-loaded conduits exhibited the ability to enhance the recovery of the injured sciatic nerve [56].

In 2023, for the first time, Puhl et al. delivered mRNA encoding NT-3 from the aligned electrospun PLLA fibers for peripheral nerve regeneration [57]. In this approach, the PLLA fiber's surface was functionalized with poly(3,4-dihydroxy-L-phenylalanine) (pDOPA) or coated with dextran sulfate sodium salt (DSS). Then, lipoplexes were obtained by complexing the synthesized pseudouridine-5'-triphosphate (Ψ)-modified mRNA encoding NT-3 (ΨNT-3mRNA) with the cationic delivery vehicle JetMESSENGER® and immobilized to the fiber PLLA surface [57]. The cationic lipoplexes containing ΨNT-3mRNA complexed to JetMESSENGER® were immobilized into PLLA fibers resulting in detectable ΨNT-3mRNA release for 21 days. Moreover, the ΨNT-3mRNA/JetMESSENGER® lipoplex-immobilized pDOPA-coated aligned electrospun PLLA fibers showed the capability to support Schwann cell secretion of NT-3 and enhance neurite outgrowth from dorsal root ganglia (DRG) neurons [57].

4.3 Periodontal tissue regeneration

Periodontal tissue is a complex tooth-supporting connective structure, in which soft, mineralized connective and epithelial tissues are structured to form a dentogingival junction [8]. However, it is highly susceptible to chronic periodontal diseases, like periodontitis, resulting in irreversible tissue destruction. Thus, to restore tissue integrity and improve periodontal regeneration, researchers have produced diverse nanofibrous membranes via electrospinning and incorporated different drugs with antibacterial, anti-inflammatory, and tissue regeneration properties [8].

Ranjbar-Mohammadi et al. prepared a PLGA/Gum tragacanth (GT) blend via blending electrospinning and PLGA/GT core-shell nanofibers via coaxial electrospinning at various ratios [58]. The tetracycline hydrochloride (TCH), a hydrophilic model drug, was incorporated within these nanofibers in order to investigate the drug-release kinetics displayed by the produced membranes. The authors obtained uniform nanofibers of PLGA, blend PLGA/GT, and core-shell PLGA/GT [58]. Moreover, the drug release of the TCH from the blend nanofibers and the core-shell structures can be controlled throughout both membranes. However, the incorporation of the TCH into PLGA/GT core-shell nanofibers exhibited a more prolonged release for 75 days with a smaller burst release within the first 2 h [58]. Thus, the more sustained release of the TCH from the core-shell membranes allied with their proven biocompatibility, antibacterial, and excellent mechanical properties make them suitable for periodontal regeneration purposes [58].

In turn, Chachlioutaki et al. used silk sericin, a natural protein derived from silkworm cocoons, and PLGA to incorporate an anti-inflammatory drug, ketoprofen, by blending electrospinning [59]. Their results revealed an increase in the hydrophilic character of the composite membranes, good mechanical properties, and a sustaining drug release for up to 15 days [59]. Besides, the in vitro assays demonstrated that the silk sericin-PLGA composite membranes promote the attachment and proliferation of human gingival fibroblasts and induce a significant downregulation of the specific pro-inflammatory markers MMP-9 and MMP-3 and an upregulation of the anti-inflammatory gene IL-10 on lipopolysaccharide-simulated RAW 264.7 macrophages [59]. Further, Zupančič et al. incorporated resveratrol (RSV) at different quantities (1, 5, 10, and 20% (w/w)) into PCL electrospun nanofibers. RSV loaded into the PCL nanofibers below 5% resulted mostly in the form of solid dispersion, while at higher loading were observed nanocrystals on the fibers surface [60]. The PCL nanofibers loaded with RSV showed a bi-phase release kinetic due to the drug dissolution and cleavage of hydrogen bonding and hydrophobic interactions between the PCL and the RSV. Thus, the RSV-loaded PCL nanofibers showed their suitability to be used as drug-delivery systems for the treatment of periodontal disease [60].

In 2022, Zhong et al. produced via electrospinning a bi-layered nanofibrous membrane composed of an antibacterial layer of PLGA/Gelatin loaded with nano-silver (nAg) and an osteoconductive layer of PLGA/Gelatin loaded with nano-hydroxyapatite (nHA) for periodontal tissue regeneration and reestablishment [61]. Both nAg and nHA were successfully incorporated into the electrospun PLGA/Gelatin nanofibers. Additionally, in vitro assays indicated that the nanofibrous membranes displayed excellent cytocompatibility and the PLGA/Gelatin nHA-PLGA/Gelatin osteoconductive layer exhibited an improved osteogenic ability for human osteoblast-like cells (MG63), as confirmed by the good cell viability and the increased alkaline phosphatase (ALP) activity, respectively [61]. On the other hand, the nAg-PLGA/Gelatin antimicrobial layer of the bi-layered nanofibrous membranes presented an effective antimicrobial capability against *S. aureus* and *E. coli* [61]. Jenvoraphot et al. combined poly(l-lactide-co-ε-caprolactone) (PLLCL) with Tetracycline to produce electrospun membranes that promote periodontal regeneration and deliver an anti-inflammatory and antibiotic drug [62]. The release profile of TC from the electrospun PLLCL membranes was characterized by a controllable slow release of the drug from the nanofibers. Besides, the in vitro assays demonstrated that the electrospun PLLCL membranes containing TC promote the proliferation of human oral fibroblast (HOF) and human oral keratinocyte (HOK) cells and exhibit antibacterial properties [62].

Peng et al. produced coaxial magnesium oxide (MgO) NPs-incorporated PCL/Gelatin core-shell nanocellulose membranes for periodontal tissue regeneration [63]. The encapsulation of the MgO within the core-shell PCL/Gelatin nanocellulose nanofibrous membranes assured their sustained release and promoted the adhesion and proliferation of human periodontal ligament stem cells (hPDLSCs) due to the biocompatibility and hydrophilicity of the Gelatin shell layer [63]. Furthermore, in vitro osteogenic and antibacterial activities were enhanced in the PCL/Gelatin nanocellulose membranes containing MgO nanoparticles [63]. Additionally, Pouroutzidou et al. reported the incorporation of the Moxifloxacin-loaded Silica-based mesoporous nanocarriers (MSNs) in PLGA composite membranes produced through electrospinning [64]. The produced drug-loaded composite fibrous membranes presented a controlled and prolonged release profile, ensuring a desirable antibacterial activity against a wide range of periodontal pathogens, good hemolytic behavior, and a protective effect on the erythrocytes [64].

4.4 Vascular grafts tissue regeneration

The electrospun nanofibers have emerged as a promising alternative to produce vascular grafts due to their ability to control the mechanical properties, the fiber diameters, the porosity, and the pore size [16]. In addition, electrospinning is a highly versatile method and allows to produce of materials with different structures and forms, as well as combining the advantages of synthetic and natural materials, which are vital for tissue-engineered grafts. Moreover, the different electrospinning techniques offer control over the alignment of the nanofibers, which can result in the orientation of the cells in a specific direction, able to provide the anisotropy encountered in certain organs including blood vessels [16].

Yang et al. prepared PCL solutions containing rapamycin (RM) that were electrospun outside the decellularized vascular graft in order to produce an RM-loaded vascular graft [65]. A small-diameter RM-loaded vascular graft showed superior mechanical properties and a sustained drug-release profile, which is considered ideal for prolonged bioactivity. Moreover, in vivo assays demonstrated that, 12 weeks after implantation, the hybrid electrospun RM-loaded decellularized vascular grafts were able to significantly reduce intimal hyperplasia without impairing reendothelialization and M2 macrophage polarization [65]. Kuang et al. combined the electrospinning and freeze-drying technology to create core-shell structures of poly(l-lactide-co-caprolactone) (PLCL) (core) nanofibers coated with heparin/silk gel (shell) [66]. The core PLCL nanofibers were designed to provide mechanical support during vascular reconstruction, while the shell heparin/silk gel layer improved the biocompatibility of the grafts. The release of heparin from silk gel-PLCL composite nanofiber graft could regulate the microenvironment in the early stage after transplantation and inhibit intimal proliferation [66]. Besides, the graft showed an auspicious biodegradation rate and was safe. Furthermore, the composite nanofiber graft provided to be a useful model for remodeling vascular structures, and in vivo assays demonstrated that the graft remained unobstructed for a long period of time (for more than 8 months) [66].

In 2021, Han et al. produced coaxial electrospun core-shell structure fibers for small-diameter vascular grafts incorporated with puerarin (PUE) [67]. The gelatin by taking PLLA was used as a carrier material, while multi-walled carbon nanotubes (MWNTs) were used as reinforced materials. The obtained results revealed that the produced Gelatin/PLLA/MWNTs-PUE fiber membranes displayed wettability, degradability, and mechanical properties similar to those presented by the natural blood vessels [67]. Besides, the PUE release from the Gelatin/PLLA/MWNTs grafts occurred in an effectively prolonged and highly efficient manner. Moreover, the fiber membrane did not induce any cytotoxic effects or side effects on endothelial cells, as well as any hemolytic activity [67]. Likewise, Maleki et al. explored the electrospinning and freeze-drying methods to fabricate a bilayer vascular graft with a core-shell structure [68]. An anticoagulant core layer composed of electrospun nanofibers of silk fibroin (SF) and thermoplastic polyurethane (TPU) containing Heparin (Hep) and a shell highly porous hydrogel layer fabricated by the freeze-drying method was used to produce a bilayer tubular vascular graft [68]. Their results revealed that the electrospun TPU-SF-Hep fibers showed robust mechanical properties, while the hydrogel layer increased the viability of the smooth muscle cells (SMCs) [68]. The Hep displayed a sustained release from the proposed graft over 40 days, as well as excellent cell and blood compatibility, as shown by 3-(4,5-dimethylthiazol- 2-yl)-2,5-diphenyltetrazolium bromide (MTT) assay and platelet adhesion test [68].

4.5 Wound dressings

The electrospun nanofibrous membranes have been widely used for efficiently wound healing and skin regeneration due to their ability to mimic the randomly oriented 3D structure of the collagen fibers found within the skin's ECM and suitable mechanical properties [8, 16]. Moreover, these membranes display high porosity, small and interconnected pore size, and high surface area, which provide a suitable microenvironment for cell adhesion and proliferation, as well as excellent nutrients/oxygen permeability, controllable evaporative water loss, and fluid drainage ability [8, 16]. Furthermore, various drugs, including antimicrobials, anti-inflammatories, growth factors, vitamins, and even cells, have been loaded into electrospun nanofibers for achieving antibacterial, antifungal, anti-inflammatory, antioxidant, analgesic, and anesthetic properties, which are vital for wound dressing applications [8, 16].

Yao et al. prepared peanut protein isolate (PPI)/PLLA nanofibers loaded with f 5, 10, and 15% of tetracycline hydrochloride (TCH) for being applied as wound dressing materials [69]. The PPI was developed using the alkali-dissolved acid precipitation method, and then the blend PPI/PLLA containing TCH was obtained via electrospinning. The incorporation of TCH did not affect the morphology of the nanofiber membranes, although the fiber diameter decreased when TCH content increased, while the wettability and mechanical properties improved [69]. Besides, the TCH was released from the PPI/PPLA composite nanofibers through a Fickian diffusion process. Moreover, the hemolysis rates were below 5%, which indicated the safety of the drug-loaded PPI/PLLA nanofibers, and the TCH-loaded PPI/PLLA composite nanofibers maintained the anticoagulant effect and showed low toxicity to cells [69]. The TCH/PPI/PLLA nanofibers also exhibited an inhibitory effect on *S. aureus* and *E. coli* growth and promoted skin wound healing in mice [69]. In turn, Yang et al. produced electrospun Janus nanofibers composed of PVP and EC loaded with a ciprofloxacin (CIP) and silver NPs (AgNPs) via a side-by-side electrospinning process as an effective antibacterial wound dressing [70]. The nanofibers presented a uniform appearance and cylindrical morphology with a clear Janus structure. The AgNPs were distributed on the EC side, while the drug was present in the PVP fibers [70]. In vitro assays demonstrated that over 90% of CIP was released within the first 30 minutes, confirming a strong antibacterial effect against both *S. aureus* and *E. coli* at the initial stages of the wound-healing process [70]. Lan et al. designed core-shell PVA/PCL nanofibers through coaxial electrospinning with dual release of tea polyphenols (TP) and ε-poly (L-lysine) (ε-PL) as antioxidant and antibacterial wound dressing materials [71]. The antioxidant TP incorporated in the PVA core showed a sustained release profile, while the antibacterial ε-PL in the PCL shell presented a fast release, allowing to achieve an antibacterial effect toward *S. aureus* and *E. coli* in the initial phase of the healing process and prolonged antioxidant activity [71]. Furthermore, the prepared coaxial core-shell nanofibers simultaneously incorporated with ε-PL and TP presented excellent cytocompatibility [71].

Zhong et al. fabricated through coaxial electrospinning PVA/PLA nanofibers embedded with *Bletilla striata polysaccharide* (BSP) and *Rosmarinic acid* (RA) to promote the wound-healing process [72]. Their results showed that the core-shell RA-BSP-PVA@PLA membranes were able to control the water vapor transmission rate (WVTR), and exhibited excellent flexibility, as well as better accommodate wounds [72]. Moreover, the MTT assay revealed that the RA-BSP-PVA@PLA nanofibers exhibited good biocompatibility and safety properties, as well as induced wound tissue growth in rat dorsal skin wound models and tissue sections [72]. Furthermore, the histological examination showed that the produced nanofibers enhanced the collagen deposition, the epidermal thickness, and the granulation tissue formation, as well as promoting the conversion of M1 macrophages into M2 macrophages, reducing the release of inflammatory factors, and promoting the occurrence of an effective wound-healing process [72].

Moreover, Mouro et al. incorporated *Chelidonium majus* L. into electrospun nanofibrous membranes composed of PCL, PVA, and Pectin (PEC) through emulsion electrospinning [73]. The electrospun PCL/PVA/PEC membranes exhibited suitable morphological, chemical, physical, and mechanical features for application as wound dressing materials [73]. Moreover, these membranes displayed excellent biological properties, namely antibacterial activity against *S. aureus* and *P. aeruginosa,* and the cytocompatibility assay provided no evidence of cytotoxicity in normal human dermal fibroblasts (NHDF cells) [73].

Recently, Sarviya et al. evaluated the performance of an ultrafine three-layer polymer nanofiber membrane produced via electrospinning [74]. In this approach, the first layer was produced with Polystyrene (PS) to act as a carrier layer and confer a suitable mechanical strength, the second layer consisted of PCL containing AgNPs endowing it with antibacterial properties, and finally, the third layer comprised of PEO act as a nonadhesive hydrophilic barrier layer with the potential to further support the healing process [74]. The cumulative Ag^+ release was improved for a period up to 84 days and the findings highlight the importance to balance the antibacterial properties with the low toxicity in order to produce a biocompatible three-layer wound dressing material containing antibacterial properties [74].

4.6 Other textile-based applications

The electrospinning technique has allowed the creation of different textile materials with controllable diameters by changing several parameters, such as the solution properties, processing conditions, and environmental variables [8]. Besides, electrospun nonwoven membranes have displayed remarkable features, like high porosity, small pore size, high surface area, and good interconnected pore structure, which permits to efficiently capture ultrafine particles, like particulate and microbial contaminants [75, 76]. Moreover, these nonwoven structures are flexible, breathable, and comfortable, and therefore electrospinning has become a promising technique for producing PPE, particularly face masks [13, 75, 77].

In 2022, Geetha et al. dispersed Zinc Oxide (ZnO) NPs into a PVA/PVP polymer blend solution through electrospinning to produce efficient antimicrobial face masks [75]. The results obtained revealed that the produced ZnO/ PVA/PVP composite nanofiber displayed a pronounced antibacterial effect against different pathogenic bacterial strains [75]. In turn, Salam et al. incorporated different concentrations of Viroblock (VB) (0.5, 1.5, 2.5, 3.5, and 5.0%) into a Polyacrylonitrile (PAN)/ZnO solution via electrospinning [78]. The electrospun nanocomposite reached an antibacterial efficiency of 92.59% against *S. aureus* and 88.64% against *P. aeruginosa*, when the higher amount of VB (5.0%) was incorporated in the nanofibers, as well as a significant reduction in virus titer (37.0%) [78]. Hence, the VB-loaded PAN/ZnO nanofibers showed the potential to develop nanofiber-based personal protective equipment, such as facemasks and surgical gowns, due to their ability to kill enveloped viruses, such as coronaviruses and influenzas [78]. Further, Ferreira et al. produced an antibacterial face-mask filter composed of PCL combined with MgO and CuO NPs using an electrospinning technique [76]. The PCL filter dopped with CuO/MgO NPs exhibited structural stability up to 2 h of washing, offered filtering capacity, and additional antibacterial activity against two different bacteria strains, *E. coli* and *S. aureus* [76].

More recently, researchers started to explore the use of electrospun nanofibers as promise carriers for the local delivery of anticancer agents in order to circumvent some of the limitations presented by the simple nanostructures, like the low stability and the burst drug-release features, and effectively prevent cancer progression and improve their therapeutic index [79]. In fact, electrospun nanofibers have many advantages in cancer treatment due to their nanofibers size, surface modification, alignment variation, and drug molecules incorporation.

In this sense, Ahmady et al. investigated the release profile of capsaicin-loaded alginate (ALG) NPs from the PCL/chitosan (CS) nanofibers. Firstly, ALG NPs were prepared using different concentrations of cationic gemini surfactant and nanoemulsions as templates. Then, the optimized ALG NPs were loaded with 20 wt% capsaicin, a pharmacological natural agent with potent anticancer activity, into a blend of PCL and CS prepared at 2:1 volume ratio. The Cap-ALG NPs @ PCL/CS nanofibers were fabricated at a constant flow rate of 0.2 mL/h, using a working distance from the needle to the rotatory drum collector of 14 cm, with a rotation speed of 300 rpm, and a voltage of 16 kV [79]. The results obtained revealed a prolonged capsaicin release from 120 h to more than 500 h when capsaicin-loaded ALG NPs were incorporated into PCL/CS nanofibers. In addition, in vitro assays also demonstrated that The Cap-ALG NPs @ PCL/CS nanofibers could effectively inhibit the proliferation of MCF-7 human breast cells and did elicit any cytotoxicity effect on human dermal fibroblasts (HDF). Therefore, the long-term and controlled release of capsaicin from the electrospun nanofibers has a high potential for the prevention and treatment of cancer [79]. In turn, Seyhan et al. prepared PLA/ Polyethylene glycol (PEG) nanofibers loaded with amygdalin (AMG) and bitter almond kernel extract in order to prevent local breast cancer recurrence [80]. The produced electrospun PLA/PEG nanofibers containing AMG exhibited a sustained and controlled release extending up to 10 h. Additionally, the acquired in vitro data revealed that the nanofibers were able to induce cytotoxicity against MCF-7 breast cancer cells [80]. Therefore, AMG-loaded nanofibers arise as a highly promising approach for reducing the risk of local recurrence of cancer after surgery and can be directly embedded into solid tumor cells for treatment. Thus, electrospun nanofibers, particularly polymeric nanofibers, can be used to incorporate antitumor drugs for sustained and controlled release of chemotherapeutic compounds at a particular site over a specific period of time with low risks of toxicity and side effects to the healthy cells. Besides, the electrospun nanofibers have received increasing consideration as promising implantable textile-based materials for the on-site delivery of chemotherapeutic drugs in a sustainable release manner after surgical resection to inhibit tumor recurrence and prolong drug release at the tumor site [81].

5. Conclusions and future outlook

Medical textiles are a field within the textile industry that has received increasing attention from researchers in recent years. Besides, the textile industry can benefit from nanotechnology, since it provides better functionalities and properties to the materials. Regarding that, tremendous efforts have been made to produce drug loading and delivery systems for tissue regeneration and wound healing able to release the drug in a controlled manner over the desired period of time.

Among the diverse methods used until now to manufacture nanoscale materials, electrospinning has been recognized as a viable technique to fabricate nanofibers from various types of textile materials that can act as carriers for the delivery of drugs due to their intrinsic properties, such as high surface area, high porosity, and small pore size. Moreover, the electrospun nanofibers have been loaded with a wide range of drugs using different electrospinning techniques, namely through blending, side-by-side, coaxial, tri-axial, emulsion, multi-needle, and needleless electrospinning. However, although the electrospinning techniques have the inherent benefits of simplicity, low cost, and versatility, all the techniques exhibit advantages and disadvantages depending on the polymeric solution and desired drug-release behavior. In fact, although the electrospun nanofibers produced by blending electrospinning are the most researched and applied for loading drugs by direct dissolving or dispersing in the polymeric solution; usually a rapid drug-release rate is observed. Thus, core–shell nanofibers produced by coaxial electrospinning and emulsion electrospinning have been highlighted to slow down the drug release. In addition, the coaxial and emulsion electrospinning provides an additional advantage for encapsulating fragile drug molecules, like growth factors, enzymes, and DNA, into the core protected by a shell, avoiding their denaturation. Further, these electrospinning techniques can involve the simultaneous use of two immiscible polymer solutions. Nonetheless, coaxial electrospinning needs a special syringe tip, while emulsion electrospinning requires the same basic setup as blending electrospinning. Moreover, side-by-side, tri-axial, and multi-needle electrospinning techniques offer the advantage of providing richer functionality from the different polymers and may result in biphasic and prolonged sustained release profiles. In turn, needleless electrospinning techniques addressed the limitations of traditional needle-based electrospinning and can produce high-throughput nanofibers using a wide variety of spinneret shapes and methods.

In addition, it is possible to tailor the properties of the electrospun textile materials by optimizing the parameters that influence the electrospinning (properties of the solution, processing variables, and environmental conditions), as well as the materials and configuration used in order to obtain the textile nanofibers with the desired morphology, size, orientation, porosity, wettability, mechanical properties, and degradation and drug-release kinetics for biomedical applications, including for bone, nerve, periodontal, and vascular tissue regeneration, wound dressings, and other textile-based materials, like PPE and cancer treatment.

However, until now, the data available concerning the nanofibers' applications at the industrial scale are still very poor, being almost limited to the laboratory scale. Thus, preclinical and clinical assays are required for further validation of these materials and to help their transition to the market. Another future outlook with the potential to revolutionize various fields could be the application of electrospun nanofibers that exhibit changes in their physicochemical properties in order to respond to environmental stimuli, where the incorporation of drugs into smart electrospun nanofibers will enable precise activation-modulated or feedback-regulated control of the drug release. As research into this technology continues to progress, the potential applications for textile electrospun nanofibers will likely expand, ranging from drug loading and delivery systems to tissue regeneration and beyond.

Acknowledgements

Financial support was provided by the Portuguese Foundation for Science and Technology (FCT), I.P./MCTES through national funds (PIDDAC), in the scope of the FibEnTech Research Unit project (UIDB/00195/2020).

Author contribution

Invited author for book contribution—Isabel C. Gouveia; Pedagogical content and methodology—Isabel C. Gouveia and Cláudia Mouro; Scientific supervision and editing—Isabel C. Gouveia; Figures and tables layout—Cláudia Mouro.

Conflict of interest

The authors declare no conflict of interest.

Author details

Isabel C. Gouveia* and Cláudia Mouro
FibEnTech Research Unit, Faculty of Engineering, University of Beira Interior, Covilhã, Portugal

*Address all correspondence to: igouveia@ubi

References

[1] Kargari A, Matsuura T, Shirazi MMMA. Electrospun and Nanofibrous Membranes: Principles and Applications. Elsevier; 2022

[2] Banerjee S, Bairagi S, Banerjee S, Ali SW, Naik SN. Recent advances in nanobiotechnology for medical textiles. In: Advances in Healthcare and Protective Text. The Textile Institute Book Series. 2023. pp. 151-171

[3] Farhaj S, Conway BR, Ghori MU. Nanofibres in drug delivery applications. Fibers. 2023;**11**:21

[4] El-Ghazali S, Khatri M, Kobayashi S, Kim IS. An overview of medical textile materials. In: Medical Textiles from Natural Resources. The Textile Institute Book Series. 2022. pp. 3-42

[5] Rostamitabar M, Abdelgawad AM, Jockenhoevel S, Ghazanfari S. Drug-eluting medical textiles: From fiber production and textile fabrication to drug loading and delivery. Macromolecular Bioscience. 2021;**21**:1-30

[6] Bairagi S, Banerjee S, Banerjee S, Wazed AS. Textile structures for medical applications. In: Advances in Healthcare and Protective Text. The Textile Institute Book Series. 2023. pp. 1-21

[7] Mehata AK, Dehari D, Vikas PV, Muthu MS. Drug-releasing textile materials: Current developments and future perspectives. In: Fiber and Textile Engineering in Drug Delivery Systems. The Textile Institute Book Series. 2023. pp. 1-38

[8] Jiffrin R, Razak SIA, Jamaludin MI, Hamzah ASA, Mazian MA, Jaya MAT, et al. Electrospun nanofiber composites for drug delivery: A review on current progresses. Polymers. 2022;**14**(18):3725

[9] Liu L, Xu W, Ding Y, Agarwal S, Greiner A, Duan G. A review of smart electrospun fibers toward textiles. Composites Communications. 2020;**22**: 100506

[10] Poshina D, Otsuka I. Electrospun Polysaccharidic textiles for biomedical applications. Text. 2021;**1**:152-169

[11] Luraghi A, Peri F, Moroni L. Electrospinning for drug delivery applications: A review. Journal of Controlled Release. 2021;**334**:463-484

[12] Miguel SP, Figueira DR, Simões D, Ribeiro MP, Coutinho P, Ferreira P, et al. Electrospun polymeric nanofibres as wound dressings: A review. Colloids and Surfaces B: Biointerfaces. 2018;**169**:60-71

[13] Xue J, Wu T, Dai Y, Xia Y. Electrospinning and electrospun nanofibers: Methods, materials, and applications. Chemical reviews. 2019;**119**: 5298

[14] Wang Y, Yu D-G, Liu Y, Liu Y-N, Wang Y, Yu D-G, et al. Progress of electrospun Nanofibrous carriers for modifications to drug release profiles. Journal of Functional Biomaterials. 2022; **13**:289

[15] Xing J, Zhang M, Liu X, Wang C, Xu N, Xing D. Multi-material electrospinning: From methods to biomedical applications. Materials Today Bio. 2023;**21**:100710

[16] Song W, Mitchell GR, Burugapalli K. Electrospinning for medical applications. Electrospinning: Principles, Practice and Possibilities. 2015:214-252

[17] Fattahi F, ZAMANI T. Release mechanisms for profen-loaded nanofibers: Challenges and

opportunities. Journal of advanced Biomedical and Pharmaceutical Sciences. 2021;**4**:195-213

[18] Kumar Sharma G, Rachel James N. Electrospinning: The Technique and Applications. IntechOpen; 2023. DOI: 10.5772/intechopen.105804

[19] Zare M, Ramakrishna S. Current Progress of electrospun nanocarriers for drug delivery applications. Proceedings. 2020;**4**:1-17

[20] Begum H, Khan MKR. Study on the various types of needle based and needleless electrospinning system for nanofiber production. International Journal of Textile Science. 2017;**6**: 110-117

[21] Mouro C, Simões M, Gouveia IC, Xu B. Emulsion electrospun fiber Mats of PCL/PVA/chitosan and eugenol for wound dressing applications. Advances in Polymer Technology. 2019;**2019**: 9859506

[22] Li J, Fu R, Li L, Yang G, Ding S, Zhong Z, et al. Co-delivery of dexamethasone and green tea polyphenols using electrospun ultrafine fibers for effective treatment of keloid. Pharmaceutical Research. 2014;**31**: 1632-1643

[23] Li Y, Wang Y, Zhang L, Ding X, Liu T, Qiu Q, et al. Electrospun silk fibroin/polyethylene oxide composite scaffolds with strontium or copper-doped hollow bioactive glass nanospheres for pH-triggered sustained drug release. Materials Letters. 2023;**336**: 133881

[24] Shi Y, Zhou M, Zhao S, Li H, Wang W, Cheng J, et al. Janus amphiphilic nanofiber membranes synergistically drive antibacterial and anti-inflammatory strategies for skin wound healing. Materials & Design. 2023;**227**:111778

[25] Li Y, Wang D, Yan E, Gao J, Wang Y, Li S, et al. Biodegradable and biocompatible Xylan/Polycaprolactone core–shell nanofibers via coaxial electrospinning for antibacterial applications. Materials Letters. 2022;**316**: 132049

[26] Ghosal K, Augustine R, Zaszczynska A, Barman M, Jain A, Hasan A, et al. Novel drug delivery systems based on triaxial electrospinning based nanofibers. Reactive and Functional Polymers. 2021;**163**:104895

[27] Ding Y, Dou C, Chang S, Xie Z, Yu DG, Liu Y, et al. Core–Shell Eudragit S100 nanofibers prepared via Triaxial electrospinning to provide a colon-targeted extended drug release. Polymers. 2020;**12**:2034

[28] Weng J, Zou Y, Zhang Y, Zhang H. Stable encapsulation of camellia oil in core–shell zein nanofibers fabricated by emulsion electrospinning. Food Chemistry. 2023;**429**:136860

[29] Varesano A, Carletto RA, Mazzuchetti G. Experimental investigations on the multi-jet electrospinning process. Journal of Materials Processing Technology. 2009; **209**:5178-5185

[30] Yoon JW, Park Y, Kim J, Park CH. Multi-jet electrospinning of polystyrene/ polyamide 6 blend: Thermal and mechanical properties. Fashion and Textiles. 2017;**4**:1-12

[31] Mouro C, Gomes AP, Gouveia IC. Emulsion electrospinning of PLLA/PVA/ chitosan with Hypericum perforatum L. as an antibacterial Nanofibrous wound dressing. Gels. 2023;**9**(9):353

[32] Mehta P, Rasekh M, Patel M, Onaiwu E, Nazari K, Kucuk I, et al. Recent applications of electrical, centrifugal, and pressurised emerging technologies for fibrous structure engineering in drug delivery, regenerative medicine and theranostics. Advanced Drug Delivery Reviews. 2021; **175**:113823

[33] Bachs-Herrera A, Yousefzade O, Del Valle LJ, Puiggali J. Melt electrospinning of polymers: Blends, nanocomposites, additives and applications. Applied Sciences. 2021;**11**:1-39

[34] Ajalloueian F, Asgari S, Guerra PR, Chamorro CI, Ilchenco O, Piqueras S, et al. Amoxicillin-loaded multilayer pullulan-based nanofibers maintain long-term antibacterial properties with tunable release profile for topical skin delivery applications. International Journal of Biological Macromolecules. 2022;**215**:413-423

[35] Nezamoleslami S, Fattahi A, Nemati H, Bagrezaie F, Pourmanouchehri Z, Kiaie SH. Electrospun sandwich-structured of polycaprolactone/gelatin-based nanofibers with controlled release of ceftazidime for wound dressing. International Journal of Biological Macromolecules. 2023;**236**:123819

[36] Mohammadzadeh V, Mahmoudi E, Ramezani S, Navaeian M, Taheri RA, Ghorbani M. Design of a novel tannic acid enriched hemostatic wound dressing based on electrospun polyamide-6/hydroxyethyl cellulose nanofibers. Journal of Drug Delivery Science and Technology. 2023;**86**:104625

[37] Pandey G, Pandey P, Arya DK, Kanaujiya S, Deepak Kapoor D, Gupta RK, et al. Multilayered nanofibrous scaffold of polyvinyl alcohol/gelatin/poly (lactic-co-glycolic acid) enriched with hemostatic/antibacterial agents for rapid acute hemostatic wound healing. International Journal of Pharmaceutics. 2023;**638**: 122918

[38] Farzaei MH, Derayat P, Pourmanouchehri Z, Kahrarian M, Samimi Z, Hajialyani M, et al. Characterization and evaluation of antibacterial and wound healing activity of naringenin-loaded polyethylene glycol/polycaprolactone electrospun nanofibers. Journal of Drug Delivery Science and Technology. 2023;**81**:104182

[39] Zech J, Leisz S, Göttel B, Syrowatka F, Greiner A, Strauss C, et al. Electrospun Nimodipine-loaded fibers for nerve regeneration: Development and in vitro performance. European Journal of Pharmaceutics and Biopharmaceutics. 2020;**151**:116-126

[40] Chen X, Ge X, Qian Y, Tang H, Song J, Qu X, et al. Electrospinning multilayered scaffolds loaded with melatonin and Fe3O4 magnetic nanoparticles for peripheral nerve regeneration. Advanced Functional Materials. 2020;**30**:2004537

[41] Rama M, Vijayalakshmi U. Synergism of silver/CEM drug on novel proteinaceous silk fibroin/mesoporous silica based sandwich-layered nanofibrous scaffolds for osteoregenerative applications. Ceramics International. 2023;**49**:17032-17051

[42] Al-Baadani MA, Hii Ru Yie K, Al-Bishari AM, Alshobi BA, Zhou Z, Fang K, et al. Co-electrospinning polycaprolactone/gelatin membrane as a tunable drug delivery system for bone tissue regeneration. Materials & Design. 2021;**209**:109962

[43] Chen CH, Chen SH, Chen SH, Chuang ADC, Darshan TG, Chen JP.

Hyaluronic acid/platelet rich plasma-infused core-shell nanofiber membrane to prevent postoperative tendon adhesion and promote tendon healing. International Journal of Biological Macromolecules. 2023:231

[44] Suresh babu J, Saravanan A, Muthuvel B, George R, Narayanan J. Synthesis and characterization of natural biomaterial composite nanofibers for ocular drug delivery systems. OpenNano. 2023;**10**:100122

[45] Hasanbegloo K, Banihashem S, Faraji Dizaji B, Bybordi S, Farrokh-Eslamlou N, Abadi PG shekhi, et al. Paclitaxel-loaded liposome-incorporated chitosan (core)/ poly(ε-caprolactone)/chitosan (shell) nanofibers for the treatment of breast cancer. International Journal of Biological Macromolecules. 2023;**230**: 123380

[46] Khatami F, Baharian A, Akbari-Birgani S, Nikfarjam N. Tubular scaffold made by gelatin/polylactic acid nanofibers for breast ductal carcinoma in situ tumor modeling. Journal of Drug Delivery Science and Technology. 2023; **85**:104606

[47] Tavira M, Mousavi-Khattat M, Shakeran Z, Zarrabi A. PCL/gelatin nanofibers embedded with doxorubicin-loaded mesoporous silica nanoparticles/ silver nanoparticles as an antibacterial and anti-melanoma cancer. International Journal of Pharmaceutics. 2023;**642**: 123162

[48] Wen M, Zhi D, Wang L, Cui C, Huang Z, Zhao Y, et al. Local delivery of dual MicroRNAs in Trilayered electrospun grafts for vascular regeneration. ACS Applied Materials & Interfaces. 2020;**12**:6863-6875

[49] Canales DA, Reyes F, Saavedra M, Peponi L, Leonés A, Palza H, et al. Electrospun fibers of poly (lactic acid) containing bioactive glass and magnesium oxide nanoparticles for bone tissue regeneration. International Journal of Biological Macromolecules. 2022;**210**: 324-336

[50] Huang C, Dong J, Zhang Y, Chai S, Wang X, Kang S, et al. Gold nanoparticles-loaded polyvinylpyrrolidone/Ethylcellulose coaxial electrospun nanofibers with enhanced osteogenic capability for bone tissue regeneration. Materials & Design. 2021;**212**:110240

[51] Alissa Alam H, Dalgic AD, Tezcaner A, Ozen C, Keskin D. A comparative study of monoaxial and coaxial PCL/gelatin/Poloxamer 188 scaffolds for bone tissue engineering. International Journal of Polymeric Materials and Polymeric Biomaterials. 2019;**69**:339-350

[52] Boraei SBA, Nourmohammadi J, Bakhshandeh B, Dehghan MM, Gholami H, Gonzalez Z, et al. Capability of core-sheath polyvinyl alcohol–polycaprolactone emulsion electrospun nanofibrous scaffolds in releasing strontium ranelate for bone regeneration. Biomedical Materials. 2021;**16**:025009

[53] Alipour H, Alizadeh A, Azarpira N, Saudi A, Alavi O, Tanideh N, et al. Incorporating fingolimod through poly(lactic-co-glycolic acid) nanoparticles in electrospun polyurethane/polycaprolactone/gelatin scaffold: An in vitro study for nerve tissue engineering. Polymers for Advanced Technologies. 2022;**33**:2589-2600

[54] Xia B, Lv Y. Dual-delivery of VEGF and NGF by emulsion electrospun nanofibrous scaffold for peripheral nerve regeneration. Materials Science and Engineering: C. 2018;**82**:253-264

[55] Alipour H, Saudi A, Mirazi H, Kazemi MH, Alavi O, Zeraatpisheh Z, et al. The effect of vitamin C-loaded electrospun Polycaprolactone/poly (glycerol Sebacate) fibers for peripheral nerve tissue engineering. Journal of Polymer Science. 2022;**30**:4763-4773

[56] Mohamadi F, Ebrahimi-Barough S, Nourani MR, Ahmadi A, Ai J. Use new poly (ε-caprolactone/collagen/NBG) nerve conduits along with NGF for promoting peripheral (sciatic) nerve regeneration in a rat. Artificial Cells, Nanomedicine, and Biotechnology. 2018; **46**:34-45

[57] Puhl DL, Funnell JL, Fink TD, Swaminathan A, Oudega M, Zha RH, et al. Electrospun fiber-mediated delivery of neurotrophin-3 mRNA for neural tissue engineering applications. Acta Biomaterialia. 2023;**155**:370-385

[58] Ranjbar-Mohammadi M, Zamani M, Prabhakaran MP, Bahrami SH, Ramakrishna S. Electrospinning of PLGA/ gum tragacanth nanofibers containing tetracycline hydrochloride for periodontal regeneration. Materials Science and Engineering: C. 2016;**58**:521-531

[59] Chachlioutaki K, Karavasili C, Adamoudi E, Bouropoulos N, Tzetzis D, Bakopoulou A, et al. Silk sericin/PLGA electrospun scaffolds with anti-inflammatory drug-eluting properties for periodontal tissue engineering. Biomaterials Advances. 2022;**133**:112723

[60] Zupančič Š, Baumgartner S, Lavrič Z, Petelin M, Kristl J. Local delivery of resveratrol using polycaprolactone nanofibers for treatment of periodontal disease. Journal of Drug Delivery Science and Technology. 2015;**30**:408-416

[61] Zhong M, Lin J, Yang Y, Liu M, Guo G, Ji D, et al. Bi-layered nanofibrous membrane with osteogenic and antibacterial functions for periodontal tissue regeneration. Journal of Biomaterials Applications. 2022;**36**: 1588-1598

[62] Jenvoraphot T, Thapsukhon B, Daranarong D, Molloy R, Supanchart C, Krisanaprakornkit S, et al. Tetracycline-loaded electrospun poly(l -lactide- co-ϵ-caprolactone) membranes for one-step periodontal treatment. ACS Applied Polymer Materials. 2022;**4**:2459-2469

[63] Peng W, Ren S, Zhang Y, Fan R, Zhou Y, Li L, et al. MgO nanoparticles-incorporated PCL/gelatin-derived coaxial electrospinning Nanocellulose membranes for periodontal tissue regeneration. Frontiers in Bioengineering and Biotechnology. 2021;**9**:216

[64] Pouroutzidou GK, Lazaridou M, Papoulia C, Tsamesidis I, Chrissafis K, Vourlias G, et al. Electrospun PLGA membranes with incorporated Moxifloxacin-loaded silica-based mesoporous Nanocarriers for periodontal regeneration. Nanomaterials. 2022;**12**:850

[65] Yang Y, Lei D, Zou H, Huang S, Yang Q, Li S, et al. Hybrid electrospun rapamycin-loaded small-diameter decellularized vascular grafts effectively inhibit intimal hyperplasia. Acta Biomaterialia. 2019;**97**:321-332

[66] Kuang H, Wang Y, Shi Y, Yao W, He X, Liu X, et al. Construction and performance evaluation of Hep/silk-PLCL composite nanofiber small-caliber artificial blood vessel graft. Biomaterials. 2020;**259**:120288

[67] Han Y, Lv Y, Chen J, Yao M, Wang C, Hu P, et al. Electrospun Core-Shell structure fibers for Puerarin-loaded vascular grafts. ACS Applied Polymer Materials. 2021

[68] Maleki S, Shamloo A, Kalantarnia F. Tubular TPU/SF nanofibers covered with chitosan-based hydrogels as small-diameter vascular grafts with enhanced mechanical properties. Scientific Reports. 2022

[69] Yao F, Zheng Y, Gao Y, Du Y, Chen F. Electrospinning of peanut protein isolate/poly-L-lactic acid nanofibers containing tetracycline hydrochloride for wound healing. Industrial Crops and Products. 2023;**194**: 116262

[70] Yang J, Wang K, Yu DG, Yang Y, Bligh SWA, Williams GR. Electrospun Janus nanofibers loaded with a drug and inorganic nanoparticles as an effective antibacterial wound dressing. Materials Science and Engineering: C. 2020;**111**: 110805

[71] Lan X, Liu Y, Wang Y, Tian F, Miao X, Wang H, et al. Coaxial electrospun PVA/PCL nanofibers with dual release of tea polyphenols and ε-poly (L-lysine) as antioxidant and antibacterial wound dressing materials. International Journal of Pharmaceutics. 2021;**601**:120525

[72] Zhong G, Qiu M, Zhang J, Jiang F, Yue X, Huang C, et al. Fabrication and characterization of PVA@PLA electrospinning nanofibers embedded with Bletilla striata polysaccharide and Rosmarinic acid to promote wound healing. International Journal of Biological Macromolecules. 2023;**234**: 123693

[73] Mouro C, Gomes AP, Ahonen M, Fangueiro R, Gouveia IC, Chelidoniummajus L. Incorporated emulsion electrospun PCL/PVA_PEC Nanofibrous meshes for antibacterial wound dressing applications. Nanomaterials. 2021;**11**:1785

[74] Sarviya N, Mahanta U, Dart A, Giri J, Deshpande AS, Khandelwal M, et al. Biocompatible and antimicrobial multilayer fibrous polymeric wound dressing with optimally embedded silver nanoparticles. Applied Surface Science. 2023;**612**:155799

[75] Geetha K, Sivasangari D, Kim HS, Murugadoss G, Kathalingam A. Electrospun nanofibrous ZnO/PVA/PVP composite films for efficient antimicrobial face masks. Ceramics International. 2022;**48**:29197-29204

[76] Ferreira CAM, Guerreiro SFC, Valente JFA, Patrício TMF, Alves N, Mateus A, et al. Advanced face mask filters based on PCL electrospun meshes Dopped with antimicrobial MgO and CuO nanoparticles. Polymers. 2022;**14**: 3329

[77] Cimini A, Imperi E, Picano A, Rossi M. Electrospun nanofibers for medical face mask with protection capabilities against viruses: State of the art and perspective for industrial scale-up. Applied Materials Today. 2023;**32**: 101833

[78] Salam A, Hassan T, Jabri T, Riaz S, Khan A, Iqbal KM, et al. Electrospun nanofiber-based Viroblock/ZnO/PAN hybrid antiviral nanocomposite for personal protective applications. Nanomaterials. 2021;**11**:2208

[79] Ahmady AR, Solouk A, Saber-Samandari S, Akbari S, Ghanbari H, Brycki BE. Capsaicin-loaded alginate nanoparticles embedded polycaprolactone-chitosan nanofibers as a controlled drug delivery nanoplatform for anticancer activity. Journal of Colloid and Interface Science. 2023;**638**:616-628

[80] Seyhan SA, Alkaya DB, Cesur S, Sahin A. Investigation of the antitumor effect on breast cancer cells of the

electrospun amygdalin-loaded poly (l-lactic acid)/poly(ethylene glycol) nanofibers. International Journal of Biological Macromolecules. 2023;**239**: 124201

[81] Razavi MS, Abdollahi A, Malek-Khatabi A, Ejarestaghi NM, Atashi A, Yousefi N, et al. Recent advances in PLGA-based nanofibers as anticancer drug delivery systems. Journal of Drug Delivery Science and Technology. 2023; **85**:104587